Henning von Nordheim

Dieter Boedeker

Jochen C. Krause

Progress in Marine Conservation in Europe

Bildautoren

Vorderseite:

Bild oben rechts: Stör Videoaufnahme (Kanada), © VOX Tierzeit 2004
Bild oben mitte: Metridium senile (Steingrund), BfN, Krause&Hübner
Bild oben links: Basstölpel (Helgoland), BfN, Krause&Hübner
Bild unten: Kegelrobbe, Franziska Tanneberger

Rückseite:

Bild oben: Flugzeug, FTZ Westküste Büsum
Bild mitte: Schiff, Kees Camphuysen
Bild unten: Taucher (Adlergrund), BfN, Krause&Hübner

Henning von Nordheim
Dieter Boedeker
Jochen C. Krause
Editors

Progress in Marine Conservation in Europe

NATURA 2000 Sites in German Offshore Waters

With 80 Figures

Springer

Dr. Henning von Nordheim
Bundesamt für Naturschutz
German Federal Agency
for Nature Conservation
BfN Insel Vilm
18581 Putbus
Germany

Dieter Boedeker
Bundesamt für Naturschutz
German Federal Agency
for Nature Conservation
BfN Insel Vilm
18581 Putbus
Germany

Dr. Jochen C. Krause
Bundesamt für Naturschutz
German Federal Agency
for Nature Conservation
BfN Insel Vilm
18581 Putbus
Germany

ISBN-10 3-642-44219-6 Springer Berlin Heidelberg New York
ISBN-13 978-3-540-33290-9 Springer Berlin Heidelberg New York

Springer is a part of Springer Science+Business Media
springer.com

Softcover re-print of the Hardcover 1st edition 2006

Cover design: E. Kirchner, Heidelberg
Production: Almas Schimmel
Printing: Krips bv, Meppel
Binding: Stürtz AG, Würzburg

Printed on acid-free paper 30/3141/as 5 4 3 2 1 0

Preface

Today, it is common knowledge that the marine environment and its biological diversity are suffering from mounting pressure caused directly or indirectly by human activities. One alarming result is the rising number of red-listed marine species and habitats worldwide, some of which are very close to extinction, while others are already extinct. Furthermore, it is not only biodiversity that is under threat; rather, many of the world's natural resources are also at risk, including our fish stocks.

For years, scientists, conservationists and non-governmental organisations have called attention to these threats, which by the beginning of the 1990s led to several global and regional international initiatives and conventions. Most prominent of these initiatives for northern European Seas are the revised Convention on the Protection of the Marine Environment of the Baltic Sea Area (Helsinki Convention) and the Convention for the Protection of the Marine Environment of the North-East Atlantic (OSPAR Convention). The Convention on Biological Diversity (CBD) should also be noted though it covers more than just the marine environment. These conventions instigated the development of long-term programmes and measures for the sustainable use of marine resources, from which emerged the definition and implementation of Ecological Quality Objectives (EcoQOs). For the marine environment, these programmes and measures were collected under the umbrella statement *Ecosystem Approach for the Management of Human Activities*, which was reaffirmed by the environment ministers of the Contracting Parties to the Helsinki and OSPAR Conventions during the first joint meeting of the Helsinki Commission (HELCOM) and the OSPAR Commission in Bremen 2003. At this ministerial meeting, an important step towards effective marine nature conservation was taken with the commitment to establish a network of "well-managed Marine Protected Areas" (MPAs) for the North-East Atlantic and the Baltic Sea by the year 2010.

Taken together with the marine NATURA 2000 sites, the resultant MPA network should provide *ecological coherence*.

NATURA 2000 is the term for the European network of protected areas within the European Union (EU). According to the legally binding European Birds and Habitats Directives, EU Member States are obliged to identify and nominate also marine NATURA 2000 sites to the European

Commission. This programme was established in 1992 by the EU Member States and the European Commission. Unfortunately, the implementation of the NATURA 2000 marine network has been hampered by issues surrounding the lack of national legal prerequisites, site selection, management, and monitoring, issues that should have been dealt with years ago. Nevertheless, the 2003 commitment of the HELCOM and OSPAR ministers to the establishment of a common MPA network by 2010 has provided a clear timeframe and a new impetus.

This book is specifically about *HabitatMareNATURA2000*, a comprehensive research programme that worked towards supplementing earlier marine assessments and supporting the implementation of NATURA 2000 in the German offshore marine area. Since the start of the programme in 2002, the German Federal Ministry for the Environment, Nature Conservation and Nuclear Safety (BMU) and its scientific advisory body, the Federal Agency for Nature Conservation (BfN), financed and conducted extensive marine research projects. The outcomes of these projects up to now, coupled with the guiding principles and rationale for the selection of marine NATURA 2000 sites, are presented in this book. The papers of the scientists involved in the programme appear here in English for the first time. This research provided essential decision support for the identification and demarcation of new MPAs in the Exclusive Economic Zone (EEZ) of the North and Baltic Seas.

As a result of the *HabitatMareNATURA2000* programme, in May 2004 ten offshore NATURA 2000 sites covering 31% of the German EEZ were nominated to the EU by BMU.

The German offshore NATURA 2000 programme is intended to also fulfil the requirements of the Helsinki and OSPAR Conventions, though this work is still ongoing. The national implementation will necessitate, apart from individual legal orders, the development of widely accepted and effective management plans, and will require the continued co-operation of the institutions and organisations which have been involved in *HabitatMareNATURA2000*.

Prof. Dr. Hartmut Vogtmann
President of the German Federal Agency for Nature Conservation

Acknowledgements

This book is the result of the joint effort of many people. Most chapters greatly benefited from the valuable comments of the following peer reviewers, whom we would like to thank for all their efforts: **Jeffrey Ardron**, Naturschutzbund Deutschland (Germany); **Kees Camphuysen**, Royal Netherlands Institute for Sea Research, Den Burg (Texel), the Netherlands; **Hans Cederwall**, Institute for System Ecology, University of Stockholm, Stockholm, Sweden; **Steven Degraer**, Gent University, Department of Biology, Gent, Belgium; **Mike Elliott**, Institute of Estuarine & Coastal Studies, Department of Biological Sciences, University of Hull, Hull, UK; **Martin Gellermann**, University of Osnabrück, Germany; **Tomas Gröhsler**, Federal Research Center for Fisheries, Institute for Baltic Sea Fisheries, Rostock, Germany; **Terry Healy**, University of Waikato, Earth Sciences Department, Hamilton, New Zealand; **Dan Laffoley**, English Nature, Petersborough, UK; **Wolfram Lemke**, Baltic Sea Research Institute, Rostock, Germany; **Mardik Leopold**, Alterra, Den Burg (Texel), the Netherlands; **Paul Nachtigall**, Hawaii Institute of Marine Biology, Kailua, Hawaii; **Daniel Owen**, Barrister, Cambridge, UK; **Jim Reid**, Joint Nature Conservation Committee, Aberdeen, Scotland, UK; **William Schwab**, US Geological Survey, Woods Hole Science Center, Woods Hole, USA; **Jonas Teilmann**, National Environmental Research Institute, Roskilde, Denmark; **Svend Tougaard**, Fishery and Maritime Museum, Esbjerg, Denmark; **Jan Warzocha**, Sea Fisheries Institute, Department of Oceanography, Gdynia, Poland; **Reinhard Zölitz-Möller**, Geographical Institute, Greifswald, Germany.

We are also grateful for the support of **Ute Hermann** and **Sandra Käning** throughout the whole process of compiling this book. In addition, special thanks are due to **Cecilia Nalagon** for the detailed copy-editing of the chapters and **Mark Parren Taylor** for the lay-out. Finally, we would like to thank particularly **Monica Verbeek**, who coordinated and was responsible for the peer-review process, final reviewing and revisions of the texts and the production of this book.

Contents

Executive Summary

Henning von Nordheim[a], Dieter Boedeker[a], Jochen C. Krause[a] and Monica Verbeek[b]

[a] *German Federal Agency for Nature Conservation, branch office: Isle of Vilm*
[b] *Aegis – Marine Environmental Consultancy, São Pedro do Estoril, Portugal*

1 Introduction

In May 2004, Germany nominated to the European Commission ten Marine Protected Areas (MPAs) in the offshore area of the German North Sea and Baltic Sea for the European network NATURA 2000. They comprise 31% of the total area of the German Exclusive Economic Zone (EEZ) and are shown in figure 1. This nomination is the result of an extensive survey programme and three years of intensive compilation. This book presents the scientific basis for the German NATURA 2000 offshore sites, and the process that lead to their selection.

2 Context

Various human activities are imposing increasing pressure on marine ecosystems. The international community has acknowledged this worrying fact by a number of global and regional initiatives, agreements, and conventions addressing marine nature conservation. Most prominent for northern European Seas are the Convention on Biological Diversity (CBD), the Helsinki Convention, the OSPAR Convention, and the EU Birds and Habitats Directives, as indicated in chapter 1. One of the most effective measures to achieve substantial marine conservation is to establish coherent networks of well managed Marine Protected Areas (MPAs) at regional and global levels. This has to be complemented by on-going single species protection, habitat enhancement, reduction of eutrophication and pollution, prevention of harmful discharges into the seas as well as reduction of pressure of resources extraction. In recent years, the above-mentioned conventions and agreements have developed programmes for the establishment of such networks of MPAs which are at present in different phases of implementation. Despite long delays, the EU Birds and Habitats Directives have been quite effective in the establishment of the NATURA 2000 network of protected marine sites in European waters, primarily in the Territorial Seas of the Member States.

The rather complex legal aspects of marine conservation through MPAs are discussed in detail in chapter 2. At the international level, under the United Nations Convention on the Law of the Sea (UNCLOS), coastal States have the competence to protect the marine environment (including through the establishment of MPAs) in their own EEZs. At the European level, there are the EU Birds and Habitats Directives, which are legally enforceable with sanctions, obliging EU Member States to protect sites in their Territorial Waters and EEZs. These sites, together, shall form a coherent ecological network of MPAs called NATURA 2000. NATURA 2000 site selection must be based on exclusively scientific criteria, wherein economic or political considerations may not play a role. Within NATURA 2000 sites, plans and projects that will or may have a negative impact on the site shall be subject of a specific environmental impact assessment procedure. In chapter 2 it is argued that EU Member States are responsible for issuing site-specific protection provisions, even if these have side effects on fisheries. In Germany, marine protection in the EEZ is implemented through Article 38 of the Federal Nature Conservation Act. This law rules out a general prohibition of certain plans and projects, thus allowing for their possibility, as long as an impact assessment is conducted, and it is safeguarded that the site's protection goals remain paramount. The German Federal Agency for Nature Conservation (BfN) is responsible for the site selection, implementation, and management of MPAs in the German EEZ.

3 Site selection procedure

In the offshore waters of the European Seas, the selection of NATURA 2000 sites has only recently begun, and guidelines for this process are still currently being developed by the European Commission. Part II of this book explains how BfN carried out the identification and selection of NATURA 2000 sites in the German EEZ as part of the *HabitatMareNATURA2000* project. The data and scientific conclusions supporting the selection are summarised in the more scientific chapters in parts III and IV of this volume. The site selection process itself was performed according to the criteria given in Annex III of the Habitats Directive and according to Article 4 of the Birds Directive.

Of the few listed natural habitat types of community interest occurring in EEZ waters and whose conservation requires the designation of NATURA 2000 sites (listed in Annex I of the Habitats Directive), only

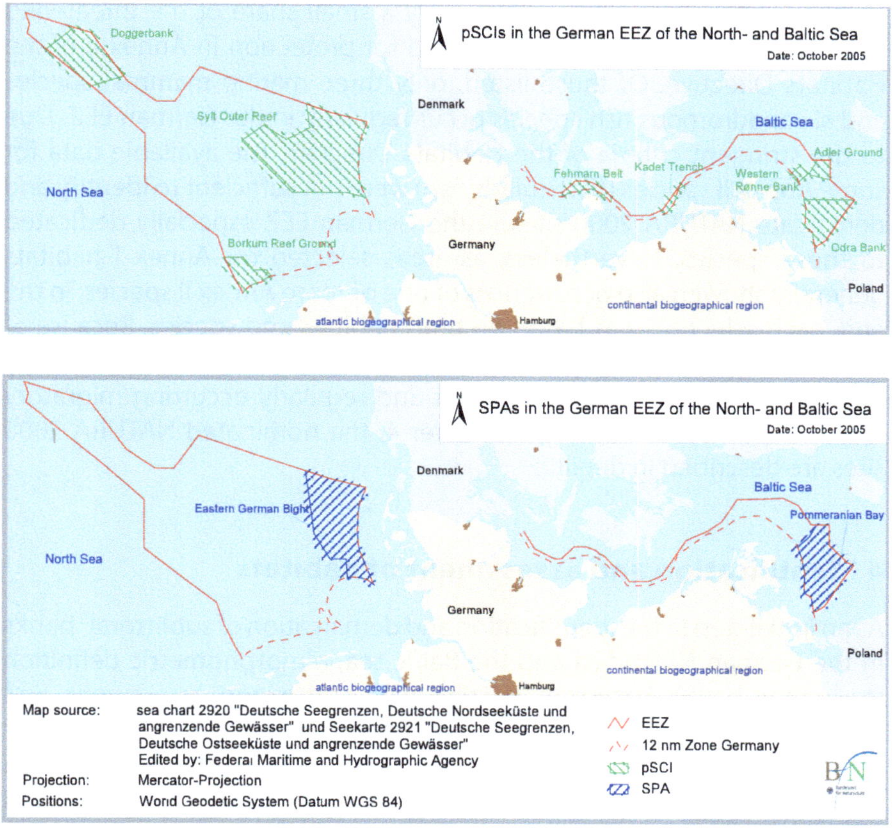

Figure 1. The ten Marine Protected Areas (MPAs) in the offshore area of the German North Sea and Baltic Sea, nominated by Germany to the European Commission for the European Network NATURA 2000: eight proposed Sites of Community Importance (pSCIs) under the Habitats Directive and two Special Protection Areas (SPAs) under the Birds Directive

sandbanks which are slightly covered by sea water all the time and *reefs* are known to occur in the German EEZ. This is discussed in chapter 3. The identification and selection of these two habitat types relied mostly upon the sediment characteristics and the plant and animal species and communities of the benthos[1]. Guided by traditional knowledge of the seafloor, data were collected using hydro-acoustic methods combined with classic benthic sampling. Several Geographical Information System (GIS) operations assisted the process of identifying reefs and sandbanks.

[1] Benthos: sea floor organisms

In chapter 4 it is pointed out that only a small share of the threatened species in the European Seas are listed for protection in Annex II of the Habitats Directive. Of those listed, only three marine mammal species and six anadromous fish species occur regularly in the German EEZ. Due to the stringent criteria of the Habitats Directive, the available data for some Annex II species (particularly fish) were not sufficient to identify and demarcate NATURA 2000 sites in the German EEZ especially dedicated to these species. Nevertheless, as areas selected for Annex I habitats generally showed also occurrences of one or more Annex II species, in the end, each selection was based on both habitats and species. Because of their long time series, the data on seabirds were sufficient for selecting sites to protect Annex I bird species and regularly occurring migratory birds of the Birds Directive. In chapter 4, the nominated NATURA 2000 sites are described in detail.

4 Identification and assessment of habitats

As an initial step in the identification and demarcation of sublittoral[2] banks in the German North Sea and the Baltic sea, a morphometric definition for marine banks, based on existing data and expert assessment, was developed and applied to a compiled set of available bathymetric data. This was aided by TIN-modelling[3] and GIS three-dimensional data analysis. As a result, 20 marine banks were identified in the North Sea, and 63 in the Baltic Sea. This information on marine banks, together with older sediment maps and existing knowledge on areas of particular ecological value, served as a basis to determine where research should be focussed. This is described further in chapter 5.

To identify reefs listed under the Habitats Directive, the banks were investigated for submarine hard-bottom substrates (e.g., layers of boulders, prominent rocks), by high-resolution seafloor imaging techniques (i.e., sidescan sonar and multibeam echosounder) as described in chapter 6.

The identification of particularly valuable benthic marine areas within the German North Sea offshore waters, including reefs and sandbanks, was based on data that were collected using grabs, trawls, underwater

2 Sublittoral: zone between the low-tide mark and the edge of the continental shelf.

3 A Triangular Irregular Network (TIN) represents the connections between the x, y, and z-coordinates of, in this case all individual bathymetric measurements. By calculating the inclinations of each triangle, a TIN model of the seafloor topography of the North Sea and the Baltic Sea could be obtained.

videos and photos in an extensive bottom-animal (zoobenthos) sampling expedition. Nine different faunal assemblages (communities) were identified, and their characteristics are presented in chapter 7. The site-specific determination of appropriate protection measures was based on the ecological values found for each area being fed into a matrix of criteria.

Four complex bank areas where rare species and rich mosaics of communities occur were selected in the North Sea for nomination for the NATURA 2000 network: *Borkum Reef Ground (Borkum-Riffgrund), Amrum Outer Ground (Amrum-Aussengrund), Sylt-Amrum Outer Reef (Sylt-Amrum-Aussenriff) and Doggerbank*. Three of these areas fulfil some or all of the ecological criteria for both sandbanks and reefs, whereas the *Doggerbank* represents an excellent example of a submarine sandbank. In addition, their distribution in space enables these banks to function as stepping-stones and refuges for migrating, fluctuating and endangered species.

For the German Baltic Sea, the benthic assessment of the four nominated NATURA 2000 sites in the offshore waters is presented in chapter 8. The *Odra Bank (Oderbank), Adler Ground (Adlergrund), Kadet Trench (Kadetrinne) and Fehmarn Belt (Fehmarnbelt)* are located along a strong salinity gradient (5–25 psu). More than 250 locations were analysed using a combination of grabs and dredges, underwater video technique, and measurements of abiotic factors (salinity, oxygen, sediment parameters). The living communities (macrophytes such as algae and seagrass, and macrozoobenthos such as worms, bivalves and crustaceans) for the different habitat types in these areas were characterised. Due to the different salinity regimes, the colonisation by benthic organisms is different. With this decreasing salinity from west to east in the Baltic, the number of marine species also declines and freshwater species increase.

5 Identification and assessment of sites – fish, mammals and birds

For the identification of sites for the six anadromous fish species listed in Annex II of the Habitats Directive, very little information was available since only a few of them are commercial species (where fishery research has traditionally tended to focus). In addition to historical data, an additional fishing study in both the North Sea and Baltic Sea was carried out to describe the distribution of Annex II-fish species in the EEZ. While chapter 9 indicates that twaite shad *(Aloxa fallax)* and river lamprey (*Lampreta fluviatilis)* are the most important Annex II-fish species in German offshore waters, no area with elevated concentrations indicating special importance to any of the Annex II-fish species have yet been found in the German EEZ of the North Sea and the Baltic Sea.

Harbour seals *(Phoca vitulina vitulina)* and harbour porpoises *(Phocoena phocoena)* are the most numerous marine mammals along the coast of the North Sea and both are listed in Annex II of the Habitats Directive. While seals are easy to study and count when they are exposed at haulouts, knowledge of seal activity at sea is still limited. New remote-sensing methods, using a satellite-compensated *dead-reckoning system* combined with a satellite tag, provided information on the utilisation of time and space by 13 of the 19 tagged harbour seals in the Wadden Sea. The analyses in chapter 10 show that the seals did not forage in the Wadden Sea, but travelled to specific hot spots in the North Sea offshore waters, far from their haulout sites in the Wadden Sea. They stayed there usually for several days, foraging on benthic prey, before turning straight back to their sandbank. It was found that seals rested not only on land (at their haulouts) but also slept in the water. In deeper water (over 10 metres), the seals rested on the surface for up to 50 minutes, apparently drifting and showing no diving activity.

To evaluate the importance of several potential NATURA 2000 sites for harbour porpoises in the German EEZ, their distribution and density were studied through aerial surveys as described in chapter 11. The main results show clear aggregations and high densities of porpoises in the areas off the North Friesian islands of Sylt and Amrum, where there are high concentrations of the species in the summer months which is their reproduction period. There seems to be a sharp gradient of density running from north to south. The highest density was consistently found in the study area *Sylt Outer Reef (Sylter Außenriff)*, followed by the *Doggerbank*. Lowest densities were calculated for *Borkum Reef Ground* . In the Baltic Sea, far fewer harbour porpoises were sighted. Their distribution showed relatively higher densities in the western part, in the *Kiel Bight* and *Flensburg Fjord (Kieler and Flensburger Bucht)*. However, in 2002 unusually high densities were observed in the eastern part of the German EEZ in the Baltic, in the study area *Pommeranian Bay (Pommersche Bucht)* on the *Odra Bank*. Such numbers were not seen again in subsequent years. Thus, the *Odra Bank* appears to constitute an occasional distribution "hot spot" for harbour porpoises for reasons yet unknown, although research is on-going.

The other two Baltic Sea study areas, *Fehmarn Belt* and *Kadet Trench,* are used by porpoises, especially the area around the island of Fehmarn. Due to the small sizes of the area, additional research methods were required. Self-contained submersible data loggers that register harbour porpoise echolocation click trains (porpoise detectors, T-PODs) were employed to carry out further research on habitat use from the island of Fehmarn in the west, to the *Pommeranian Bay* in the east. The results of this research given in chapter 12 confirm the decrease in harbour porpoise density from the west to the east in the German Baltic Sea, as found in the aerial surveys (see chapter 11).

For the selection of Special Protection Areas (SPAs) for birds under the Birds Directive as part of the NATURA 2000 network, seabird concentration areas in the German North Sea and Baltic Sea were identified by aerial and ship-based surveys (see chapter 13). Species distribution maps based on density were produced from these data. For widely distributed species, patterns were analysed by grid maps. For aggregating species, spatial interpolation (ordinary kriging) was used. For each species of relevance, concentration areas were determined and subsequently combined so that a set of candidate conservation areas could be identified, leading to the derivation of potential SPAs. In the German North Sea, the heterogeneity of the data did not allow the use of geo-statistical methods. As a result, one large single site, the *Eastern German Bight (Östliche Deutsche Bucht)* was demarcated covering the wintering, resting and feeding areas of Annex I species under the Birds Directive, which were mainly red-throated and black-throated divers (*Gavia stellata* and *G. arctica)*. In the German Baltic Sea, clusters of various seabirds, mostly marine ducks including long-tailed duck (*Clangula hyemalis)* and common scoter (*Melanitta nigra*) were much more conspicuous, and the combination of several concentration areas resulted in the selection of an area covering the *Pommeranian Bay* and the adjacent *Adler Ground*. In chapter 13, field methods and methods of analysis are briefly discussed and further recommendations are given.

For the identification of sites for seals (chapter 10), harbour porpoises (chapters 11 and 12) and birds (chapter 13), BfN made use of the results of the combined research project MINOS (Marine Mammals and Birds In the North and Baltic Seas). This project, described in chapter 14, was conducted in seven separate subprojects examining whether large-scale offshore wind farms in the German North Sea and Baltic Sea affect or endanger harbour porpoises, common seals and seabirds. The results are expected to provide better knowledge for estimating and assessing the impacts of future wind farms planned in the German EEZ. MINOS focuses on two issues: the preferential habitats and migratory routes of these species or species groups in the EEZ, and the sense of hearing and sensitivity of porpoises and seals, in order to assess possible damage, displacement and disturbance by offshore wind farms.

6 Public awareness and consultation

All relevant information on Habitats Directive Annex I and II species and habitats, as well as the distribution of seabirds according to the Birds Directive, were compiled in a Geographical Information System (GIS), on the basis of which ten sites were selected. These ten demarcated sites were presented and discussed with the relevant German federal ministries and the coastal States (Länder), and were presented in a public consultation process, as required by the German Federal Nature Conservation Act (see chapter 15). To facilitate these consultations, in November 2003 the proposed areas were listed and mapped on the Internet (www.*HabitatMareNATURA2000*.de), published in newspapers and distributed in press releases. Citizens had the opportunity to submit written comments or participate in three public hearings that were held in the coastal States in December 2003. Any substantial nature conservation data and responses from this process were used to improve the data forms and to change the shape of one site. However, many proposed amendments that reflected only socio-economic interests could not be considered since the Habitats Directive clearly states that for the selection of sites only conservation criteria can be considered. Altogether, this process took more than one year. Finally, in May 2004, the nomination of ten NATURA 2000 sites in the German EEZ was submitted to the European Commission in Brussels.

As part of the *HabitatMareNATURA2000* programme, parallel to the selection and nomination process of the marine NATURA 2000 sites, four products to increase public awareness were created and are described in chapter 16. A booklet, a video and a website, all in German and English, were launched in June 2003 at the first joint ministerial meeting of the Helsinki and OSPAR Commissions in Bremen. An interactive CD-Rom was published in the beginning of 2006.

Introduction

The aim of this publication is to present and to share recent progress made on nature conservation within German marine offshore waters with a larger, international audience. It is specifically about the outcome of the research and assessment project *HabitatMareNATURA2000*, a comprehensive programme oriented towards the implementation of a network of marine protected areas (MPAs) in the German *Exclusive Economic Zone* (EEZ), as part of NATURA 2000. The legal background for NATURA 2000 is the *Habitats Directive* of the European Union. It entered into force in 1992 and the Member States of the European Union (EU) committed themselves to designate protected areas for the conservation of natural habitats and of wild fauna and flora in Europe. Such protected areas, called *Special Areas of Conservation* (SACs), together with the *Special Protection Areas* (SPAs) designated under the EU *Birds Directive* (1979), form the NATURA 2000 network.

In Germany, for legal reasons the selection and establishment of marine NATURA 2000 sites were initially only possible within the *Territorial Waters* (up to12 nautical miles from shore). Within these waters, the German coastal States (Länder) are responsible for nature conservation. When the Federal German Nature Conservation Act was amended in April 2002, it established the statutory basis for the implementation of NATURA 2000 in the EEZ (marine area beyond the Territorial Waters, and maximum 200 nautical miles offshore).

In May 2002, *HabitatMareNATURA2000*, initiated and financially supported by the German Federal Ministry for the Environment, Nature Conservation and Nuclear Safety (BMU) and its scientific advisory body, the Federal Agency for Nature Conservation (BfN), embarked on extensive marine research regarding the identification and spatial distribution of NATURA 2000 features of interest within the EEZ.

According to both the Habitats and the Birds Directives, specific natural habitats such as reefs and sandbanks, as well as several species of marine mammals, fishes, and birds are to be protected by NATURA 2000. Earlier scientific studies had already provided some data on the occurrence and distribution of these species and habitats. The results of *HabitatMareNATURA2000* helped fill in the gaps, providing BMU and BfN with sufficient data to act upon the selection and determination of NATURA 2000 sites in the EEZ.

In May 2004, Germany nominated to the European Commission ten EEZ marine sites as its contribution towards NATURA 2000. The two SPA sites achieved in September 2005 the national legal status of nature reserves. Overall, Germany has now designated approximately 31% of its EEZ and 38% of its total marine area (when the currently nominated sites from the Territorial Seas are included) as NATURA 2000 sites.

The papers presented in this book on marine mammals and seabirds reflect assessments of the joint data sets collected by both *HabitatMareNATURA2000* and a related scientific marine research project entitled *MINOS*, which has been running in parallel. Whereas the surveys and investigations under the MINOS project comprised, as far as possible, the whole German marine area (i.e., the Territorial Waters and the EEZ), the NATURA 2000 surveys mainly concentrated on closer investigations of areas already known to be of particular ecological importance. Close cooperation between the two projects avoided unnecessary repetitions or redundancies, particularly with wide ranging species such as harbour porpoises.

This book is structured as follows:

- **Part I** provides the reader with relevant information on the international, European and national legal aspects, and relevant marine conventions and agreements;
- **Part II** describes in detail how the BfN team used available scientific data and the results of the different NATURA 2000 and related projects for the identification and demarcation of NATURA 2000 sites in the German EEZ;
- **Parts III and IV** deal with the identification of relevant habitats and species for the site selection;
- **Part V** gives a short overview of how BMU carried out inter-governmental and public consultations, and also of the tools BfN employed to promote the NATURA 2000 programme.

Three contributions in this volume were written by scientific and legal teams who were not directly involved in the NATURA 2000 programme: **Czybulka and Bosecke** contributed as jurists and were involved in clarifying many legal questions, particularly those concerning the national implementation of marine NATURA 2000; **Kellermann et al.** presented the scope and first results of the MINOS project; **Liebsch et al.** investigated and analysed movements of tracked common seals within the MINOS project, the results of which were also used by the NATURA 2000 programme.

The editors, contributors, and publisher hope that this book will be interesting and useful not only for the scientific community but also for the broader public, particularly the members of governmental and non-governmental organisations engaged in questions of marine conservation. Furthermore, it is hoped that the methodologies and rationales used in the establishment of the German offshore-MPAs would inform and inspire others involved in similar projects within the EU and abroad.

The editors

Part I: MPAs in the German EEZ – conventions and legal aspects

Chapter 1

International conventions for marine nature conservation and marine protected areas relevant to the North Sea and the Baltic Sea

Henning von Nordheim, Dieter Boedeker and Jochen C. Krause

German Federal Agency for Nature Conservation, branch office: Isle of Vilm

Abstract

Increasing pressure on marine ecosystems resulting from various kinds of human activities has led to a number of global and regional international initiatives and conventions. Most prominent for northern European Seas are the Convention on Biological Diversity (CBD), the Helsinki Convention on the Protection of the Marine Environment of the Baltic Sea, the OSPAR Convention for the Protection of the Marine Environment of the North-East Atlantic, and the Birds and Habitats Directives of the EU.

The fundamental goals of marine nature conservation can be described as to conserve, protect, restore and manage:

- the functioning of marine ecosystems and their services;
- the regenerative capacity of natural resources and their sustained availability for human use;
- natural diversity, population dynamics, density and age structure of marine fauna and flora;
- intact and diverse natural marine habitats and biotopes; and
- the characteristic features and beauty of nature and viewscapes.

These conservation objectives can only be achieved if human activities affecting our oceans and seas were conducted by applying six key principles:

(1) the precautionary principle;
(2) best environmental practices (BEP);
(3) best available technologies (BAT);
(4) the polluter-pays principle;
(5) the compensation or substitution principle; and
(6) the avoidance principle.

Further, the ecosystem approach to the management of human activities as has been clarified by the Helsinki- and the OSPAR Conventions should always be applied.

Probably the most effective measure to achieve substantial marine conservation is to establish coherent networks of Marine Protected Areas (MPAs) at regional and global levels. This needs to be complemented by general habitats and species protection and enhancement, as well as by undiminished efforts to reduce unsustainable extraction of natural resources, to minimize eutrophication and pollution and to hinder new discharges to the seas. Red lists of threatened, endangered and/or declining species, habitats or biotopes would help assess the status of marine biodiversity, the effects of human activities, and the possible success of interventions.

This paper describes the most relevant international conventions and regulations with respect to marine nature conservation of the seas in Northern Europe (North East Atlantic and Baltic Sea). The current implementation status of conservation measures and specifically of MPAs and some next steps are also outlined.

1 Introduction

Over the past decade, it has been globally recognised that the marine environment and its biodiversity are increasingly under pressure, not only due to pollution and eutrophication, but also because of over-exploitation and degradation linked to the individual and combined effects of human activities such as fisheries, shipping, tourism, and offshore wind energy projects, offshore oil and gas extractions, and offshore mining (De Fontaubert et al. 1996, HELCOM 2002, ICES 2003, OSPAR 2000).

In Northern Europe, programmes to preserve marine ecosystems such as controlling emissions, banning incineration at sea, phasing out the use of Tributyltin (TBT) and other hazardous substances, improving regulations on oil and gas exploitation at sea and advanced techniques in wastewater treatment have achieved measurable successes since their inception in the early 1980s (EEA 2001, HELCOM 2003, OSPAR 2000). However, the degradation of marine ecosystems and the continuous loss of biodiversity have not been stopped by these measures. Hence, additional new marine ecosystem protection targets and goals were needed.

The "ecosystem approach", a widely accepted concept at present, and which has its roots in the Convention on Biological Diversity, evolved to meet these new goals (SRU 2004). This approach was further developed and specified by HELCOM and OSPAR in 2003 with particular emphasis on applying the precautionary principle in the management of human activities for the protection of the marine environment of the North East Atlantic and the Baltic Sea area (JMM 2003a).[1]

Other action programmes and high-level conferences such as the World Congress on National Parks and Protected Areas held in Caracas, Venezuela in 1992 called for the establishment of a global network of MPAs as a useful tool in the protection of marine biodiversity.

After the political changes at the end of the 1980s, the need for protection of the marine environment became widely recognised and accepted. Four global and regional conventions and regulations highly relevant to improved protection of the marine environment and biodiversity of Northern European waters were established in 1992: the Convention on Biological Diversity (CBD), the ("new") Helsinki-Convention, the OSPAR Convention and the EU Habitats Directive

---

[1] The ecosystem approach to the management of human marine activities shall safeguard the balance between sustainable use and protection of the marine biological diversity. By considering the functioning of the marine ecosystem in its entirety, including humans as an important factor, it aims at maintaining the health and fitness of the marine ecosystems in harmony with an appropriate and sustainable use of marine goods and services, for the benefit of present and coming generations.
The ecosystem approach to human activities comprises the following key elements:

- Integrated consideration of the total marine ecosystem
- Management across all sectors
- Stakeholder participation
- Establishment and application of ecological quality objectives .

supplementing the EU Birds Directive of 1979. The core messages of these conventions and directives again received support from the decisions of the World Summit on Sustainable Development (WSSD) in Johannesburg (2002) and particularly from the IUCN World Park conference in Durban (2003) and the 7th CBD conference of parties in 2004 (Kuala Lumpur).

The following sections present a background on marine conservation objectives and an introduction to the main global and regional conventions and acts regarding MPAs.

2 General marine nature conservation objectives, management principles, and tools

The fundamental goals of marine nature conservation can be described as conservation, protection, management, and where necessary, restoration, in order to safeguard on a lasting basis:

- the functioning of marine ecosystems and their services;
- the regenerative capacity of natural resources and their sustained availability for human use;
- natural diversity, population dynamics, density and age structure of marine fauna and flora;
- intact and diverse natural marine habitats and biotopes; and
- the characteristic features and beauty of nature and viewscapes.

Many major threats to marine nature, including climate change, can generally be traced back to human mismanagement. Consequently, the realisation of these fundamental objectives would most benefit if human activities in our oceans and seas were to be carried out only after passing thorough scientific review and a comprehensive environmental impact assessment (EIA) and if the following **6 key principles** were applied:

(1) the precautionary principle
human activities, that according to current scientific knowledge may lead to serious negative impacts on the marine ecosystem and that may not be at all or sufficiently foreseeable at present, shall not be conducted until additional knowledge proves that is it not harmful

(2) best environmental practices (BEP)
human activities shall be carried out in a manner that safeguards in the best possible way all components of the marine ecosystems

(3) best available technologies (BAT)
human activities shall only be conducted by applying the best available technology that has the least impact on the marine environment

(4) the polluter-pays principle
It is not the society or community but the one who causes a negative impact or destruction who is responsible for its removal or who must prove that the impact is harmless

(5) compensation or substitution principle
human activities that lead to negative impacts on or destruction of parts of the ecosystem have to be balanced by compensatory measures or by substitution measures (e.g., money)

(6) avoidance principle
all human activities that lead to negative impacts on the marine ecosystem, that are unnecessary, unsustainable or cannot be compensated, shall not be carried out.

These principles and its effective implementations are complemented by the application of the ecosystem approach to the management of human activities as has been clarified by HELCOM and OSPAR (JMM 2003a) as mentioned above.

Many of these principles are addressed in relevant paragraphs in the Helsinki and the OSPAR Conventions. However, they have yet to see consequent implementation by the contracting parties.

Apart from species protection and enhancement programmes directed towards species or populations under direct threat or decline, probably the most effective measure to achieve the above-mentioned basic aims of marine conservation is to establish a coherent network of MPAs. The World Conservation Union (IUCN) defines a *marine protected area* as

> *any area of intertidal or subtidal terrain, together with its overlying water and associated fauna, flora, historical and cultural features, which has been reserved by law or other effective means to protect part or all of the enclosed environment* (Kelleher 1999).

This is the definition used throughout this publication.

MPAs can serve many different purposes and are important tools for marine nature conservation (Kelleher 1999, Salm and Clark 2000). However, ecologically coherent networks of MPAs and other programmes or measures for the achievement of the above marine nature conservation goals have to be supported with undiminished efforts to reduce eutrophication and pollution, and to hinder new direct or indirect (i.e., through rivers, air) discharges to the seas. These can go hand-in-hand with setting up so-called "ecological quality objectives".

Further, it is critical that marine resources are used sustainably, with the clear understanding that some species or habitats may not be able to withstand exploitation of any kind. Additionally, non-renewable marine resources, e.g., sand and gravel, need a wise-use approach that sets aside major portions of them for generations to come.

A very important tool to help assess marine biodiversity, the effects and consequences of human activities, as well as the success of measures to improve the marine environmental conditions are the (Red) Lists of threatened, endangered, and/or declining species and/or habitats such as those established by HELCOM (Nordheim and Boedeker 1998) and by OSPAR (OSPAR 2004).

3 International conventions and regulations relevant for marine nature conservation in Northern Europe

In the early 1990s, at least four global or regional conventions, as well as the EU Habitats Directive, relevant to the protection of the marine environment and marine biodiversity in Northern Europe, i.e., the North East Atlantic and the Baltic Sea were established (table 1).

It should be mentioned that for southern European seas (i.e., the Mediterranean and the Black Sea), the Barcelona Convention and the Bucharest Convention represent similar important regional marine conventions with regard to the protection of the Mediterranean and Black Sea, respectively.

3.1 United Nations Convention on the Law of the Sea (UNCLOS)

The United Nations Convention on the Law of the Sea (UNCLOS), the so-called *"constitution for the seas"*, established a legal regime in the world's oceans intending to govern all uses of the oceans and their resources. It entitles coastal States to determine territorial limits in the oceans and seas that are important for all ensuing regulations, including those for the environment (see chapter 2).

Part XII of the Convention is devoted solely to the marine environment and lays down the fundamental obligation of all States *"to protect and preserve the marine environment"* (Art. 192). This part of UNCLOS intends primarily to regulate and prevent ship-borne pollution. Although UNCLOS limits a coastal nation's competence to regulate foreign navigation in its EEZ, shipping issues can be addressed at the global or regional level, for example, by approaching the International

Maritime Organization (IMO). In this respect, Article 211(6)(a) and (c) entitles coastal States to apply at IMO for the establishment of PSSAs (Particularly Sensitive Sea Areas). A PSSA is a maritime area that needs special protection through action by the IMO (Gjerde 2002). Article 194(5) states that measures taken *"shall include those necessary to protect and preserve rare or fragile ecosystems as well as the habitat of depleted, threatened or endangered species and other forms of marine life."* There exist presently seven designated PSSAs: the Great Barrier Reef, Australia (1990); the Sabana-Camagüey Archipelago in Cuba (1997); Malpelo Island, Colombia (2002); the sea around the Florida Keys, United States (2002); the Wadden Sea, Denmark, Germany, Netherlands (2002); Paracas National Reserve, Peru (2003); and the Western European Waters (2004). The Marine Environment Protection Committee of IMO (MEPC) has also approved in principle the following PSSAs (designation will be considered at future sessions of the MEPC): Torres Straits (Australia and Papua New Guinea), the Baltic Sea (except Russian waters), waters of the Canary Isles (Spain), and Galapagos Archipelago (Ecuador) (IMO 2005).

UNCLOS is the most important international legal basis for an application of the ecosystem approach to the management of human activities in marine areas that are beyond the national jurisdiction of a coastal State or in marine areas with only limited national jurisdiction (i.e., EEZ). Apart from the IMO, the International Seabed Authority (ISA) and the different regional fisheries management commissions (RFMOs), such as the North East Atlantic Fisheries Commission (NEAFC), are also competent authorities for regulating human activities that may have impacts on marine biodiversity.

3.2 Convention on Biological Diversity (CBD)

Among many other issues, the CBD stresses in Article 8 in general terms that *"each Contracting Party shall, as far as possible and as appropriate... establish a system of protected areas where special measures need to be taken to conserve biological diversity."* While this article does not explicitly mention MPAs, the United Nation Conference of Environment and Development, 1992 (UNCED) in Rio stated in chapter 17 of its general programme, "Agenda 21", that coastal States should undertake measures to maintain biological diversity and productivity of marine species and habitats, including the establishment and management of protected areas. However, at that time no concrete action plan for the establishment of MPAs was agreed upon by the CBD.

Table 1. Important International Conventions and regulations for the establishment and management of MPAs in Northern Europe (including important articles (Art.), decisions (Dec.), recommendations (Rec.), agreements (Agr.))

<table>
<tr><th>Convention/ regulation</th><th>Acronym</th><th>Adoption, date</th><th>Geographical scope</th></tr>
<tr><td>United Nations Convention on the Law of the Sea</td><td>UNCLOS</td><td>Montego Bay, Jamaica
signed 1982
entered into force 1994</td><td>World oceans and seas</td></tr>
<tr><td colspan="4">Relevant components
General Obligations, Part XII (Art. 192–237):
• General Protection and Preservation of the Marine Environment (Art. 192)
• Measures to prevent, reduce and control pollution of the marine environment (Art. 194), including those necessary to protect and preserve rare or fragile ecosystems as well as the habitat of depleted, threatened or endangered species and other forms of marine life (Art. 194 (5))
• International Rules and National Legislation to Prevent, Reduce and Control Pollution of the Marine Environment (Art. 207–212)
• Enforcement by Flag States (Art. 217)
Specific Obligations
• Jurisdictions with regard to the protection and preservation of the marine environment in the EEZ (Art. 56 (1) b) iii))
• Prevention, reduction and control of pollution from pipelines on the continental shelf (Art. 79 (2))
• Conduct of marine scientific research in compliance with all relevant regulations including those of the protection and preservation of the marine environment (Art. 240 d)</td></tr>
<tr><td>Convention on Biological Diversity</td><td>CBD</td><td>Nairobi, Kenya
adopted 1992
entered into force 1993</td><td>World wide</td></tr>
<tr><td colspan="4">Relevant components and decisions
• In-situ Conservation, Protected Areas (Art. 8)
• Decision on the Conservation and Sustainable Use of Marine and Coastal Biological Diversity – Jakarta Mandate (Dec. II/10), Jakarta, Indonesia 1995
• Decision on Marine and Coastal Biological Diversity (Dec. VII/5), Kuala Lumpur, Malaysia 2004
• Decision on Protected Areas (Dec. VII/28), Kuala Lumpur, Malaysia 2004</td></tr>
<tr><td>Convention on the Protection of the Marine Environment of the Baltic Sea Area</td><td>Helsinki Convention</td><td>(first: signed 1974)
signed 1992;
entered into force in 2000</td><td>Baltic Sea (incl. Kattegat)</td></tr>
<tr><td colspan="4">Relevant components and decisions
• Nature Conservation and Biodiversity (Art. 15), 1992
• System of Coastal and Marine Baltic Sea Protected Areas (BSPA) (Rec. 15/5), 1994
• Protection of heavily endangered or immediately threatened Marine and Coastal Biotopes in the Baltic Sea Area (Rec. 21/4), 2000</td></tr>
</table>

Table 1 continued.

Convention/ regulation	Acronym	Adoption, date	Geographical scope
Convention for the Protection of the Marine Environment of the North-East Atlantic (OSPAR Convention)	OSPAR	signed 1992 entered into force in 1998	North-East Atlantic (incl. the North Sea)
Relevant components and decisions • Annex on the Protection and Conservation of the Ecosystems and Biological Diversity of the Maritime Area (Annex V), Sintra, Portugal, 1998 • Recommendation on a Network of Marine Protected Areas (Rec. 2003/3), 2003 • Initial OSPAR List of Threatened and/or Declining Species and Habitats (Agr. 2004-06), 2004			
European Coherent Network of (terrestrial & marine) Protected Areas	NATURA 2000	*Brussels, Belgium 1992*	Area of EU Member States (including EEZs)
Composed of the: **(1) Directive on the Conservation of Wild Birds** (Birds Directive) (79/409/EEC), 1979 **Relevant components:** • Classification of the most suitable territories in number and size as "Special Protection Areas" (SPA) for species listed in Annex I (Art. 4.1) • Taking of similar measures for regularly occurring migratory species not listed in Annex I (Art. 4.2) • Protection of Species (Art. 5-9) **(2) Directive on the Conservation of Natural Habitats and of Wild Fauna and Flora** (Habitats Directive) (92/43/EEC), 1992 **Relevant elements:** • Conservation of Natural Habitats and Habitats of Species (Art. 3–11) • Protection of Species (Art. 12–16) • Confirmation of the Birds and the Habitats Directive (Art. 7) • Natural Habitat Types of Community Interest whose Conservation requires the Designation of Special Areas of Conservation (SAC) (Annex I) • Animal and Plant Species of Community Interest whose Conservation requires the Designation of Special Areas of Conservation (SAC) (Annex II) • Identification of "Special Areas of Conservation" (SACs) (Annex III)			

Table 1 continued.

The following conventions are also relevant for marine areas within and beyond the Territorial seas (Gjerde 2002), but they will not be introduced here in detail:

1. Convention on the Conservation of Migratory Species of Wild Animals (Bonn Convention, 1979); and agreements under the Bonn Convention for European coastal and marine species:
- Agreement on the Conservation of Seals in the Wadden Sea - CWSS (1990)
- Agreement on the Conservation of Small Cetaceans on the Baltic and North Seas
- ASCOBANS (1991)
- Agreement on the Conservation of Cetaceans of the Mediterranean and Black Seas
- ACCOBAMS (1996).

2. Convention on the Conservation of European Wildlife and Natural Habitats (Bern Convention, 1979).

3. Relevant regional conventions for Southern European Seas:
- **Convention for the Protection of the Mediterranean Sea** against Pollution (Barcelona Convention, 1976)
- Protocol Concerning Specially Protected Areas and Biological Diversity in the Mediterranean to the Barcelona Convention (Barcelona, 1995)
- **Bucharest Convention on the Protection of Black Sea** against Pollution (Bucharest, 1992)

The *Jakarta Mandate on Marine and Coastal Biological Diversity* is such an action plan. It was adopted by the 2nd Conference of the Parties (COP II) of the CBD in 1995 and was reviewed and updated in 2004 in Kuala Lumpur. The plan focuses on integrated marine and coastal area management (IMCAM), the sustainable use of living resources, marine and coastal protected areas (MCPA), mariculture, and the introduction of non-indigenous species. In 2004, the COP VII further adopted a concrete timeframe for the realisation of the programme (CBD 2004) and set the target to develop such MPCA systems by 2012. A gap analysis of protected areas is to be completed by 2006, and by 2008 action is to be taken by the parties to address the under-representation of marine and inland water ecosystems in existing national and regional systems of protected areas (CBD 2005). Thus, the CBD has finally developed an ambitious and comprehensive programme to protect marine habitats and species through a network of MPAs.

3.3 Helsinki Convention

The first "Helsinki-Convention" of 1974 entered into force during the Cold War. It covers the maritime area of the Baltic Sea, Belt Sea, and Kattegat; this consists only of Territorial Seas and EEZs of the nine riparian states.

The Convention was very successful in assessing and combating marine pollution for coastal States that had very different political systems at that time. The environmental condition of the Baltic Sea improved as a result of several specific measures such as the ban on DDT. However, the Convention of 1974 did not specifically address protection of marine nature and biodiversity.

After the political changes at the end of the 1980s, the international discussions on environmental and biodiversity issues between Baltic nations and NGOs intensified. Conservationists strongly petitioned for new measures to protect biodiversity and thus called for an amendment of the 1974 Convention. Consequently, the "new" Helsinki Convention was signed in 1992 with 9 Member States and the EU as Contracting Parties (http://www.helcom.fi/). Article 15 provided, for the first time, in a regional sea convention, clear and wide-ranging commitments to marine nature conservation and biodiversity protection:

> *The contracting parties shall individually and jointly take all appropriate measures with respect to the Baltic Sea Area and its coastal ecosystems influenced by the Baltic Sea to conserve natural habitats and biological diversity and to protect ecological processes.*
>
> *Such measures shall also be taken in order to ensure the sustainable use of natural resources within the Baltic Sea Area. To this end, the Contracting Parties shall aim at adopting subsequent instruments containing appropriate guidelines and criteria.*

HELCOM System of Marine and Coastal Baltic Sea Protected Areas

In 1993, the Helsinki Commission (HELCOM) established an Expert Working Group on Marine Conservation and Biodiversity (EC-NATURE), chaired by the first author for 10 years. One of its first activities led to the construction in 1994 of a system of marine and coastal "Baltic Sea Protected Areas" (BSPA) (HELCOM Recommendation 15/5). Sixty-two BSPAs with preliminary site delimitations were submitted as a first round of site proposals to the Helsinki Commission (see also HELCOM 1996). Common guidelines were agreed upon, including criteria for the selection of such areas and guidelines for the establishment of management plans. The recommendations are "soft law". Details of the HELCOM Recommendation 15/5 and of the guidelines for the establishment of BSPAs are summarised in table 2.

Table 2. Details of HELCOM Recommendation 15/5 and of the guidelines for the establishment of BSPAs (2003) (http://www.helcom.fi./Recommendations/en_GB/rec15_5/)

Rec. 15/5 (System of Baltic Sea Protected Areas, BSPA) • the size of a protected area of this system should not be less than 1000 hectares; • the system shall be enlarged stepwise by additional areas, preferably purely marine (i.e., offshore), and additional coastal terrestrial areas; • an alteration of an area and its protection status, once it is notified to HELCOM, can only be conducted if HELCOM is consulted prior to such an action; • a management plan must be established for each area following the BSPA guidelines (below), taking into consideration possible negative impacts and stakeholder interests for that area; • a monitoring programme shall be conducted covering biological, chemical, and physical parameters, and representing an integrated part of the existing Baltic Sea monitoring programme of HELCOM; and • every three years the status of the establishment of this system shall be reported to HELCOM.
BSPA Guidelines (2003) • Selection of sites is conducted by applying a broad suite of identification criteria; • Only national protection will ensure a high degree of legally binding protection of designated BSPAs; • For EU Member States, management in line with the EU Habitats Directive and the EU Birds Directive can also fulfil the necessary protection and management requirements for BSPAs; and • An application for approval of a new BSPA can be sent at any time to the HELCOM Secretariat.

The situation of the final and legal establishment of BSPAs by the various Baltic Sea States is illustrated in figure 1. To date, few of the 1994 designated areas have been fitted conclusively into the system of BSPAs, that is, many of the designated areas have not yet been finally established by sending definitive maps and management plans to HELCOM (HELCOM 2002). Only Lithuania has already established all of its originally notified BSPAs (three areas). Nevertheless, several of the BSPAs hold a national protection status. As a further task of the contracting parties to HELCOM, there still remains the need to incorporate into this system at least those 24 additional offshore areas which were identified by experts in 1998 (Hägerhäll and Skov 1998). In the end, the BSPAs will be a very powerful instrument for protecting marine biodiversity of the Baltic Sea area and in particular those animals, plants and biotopes that are under threat or decline and have been identified by the Red Lists of marine and coastal

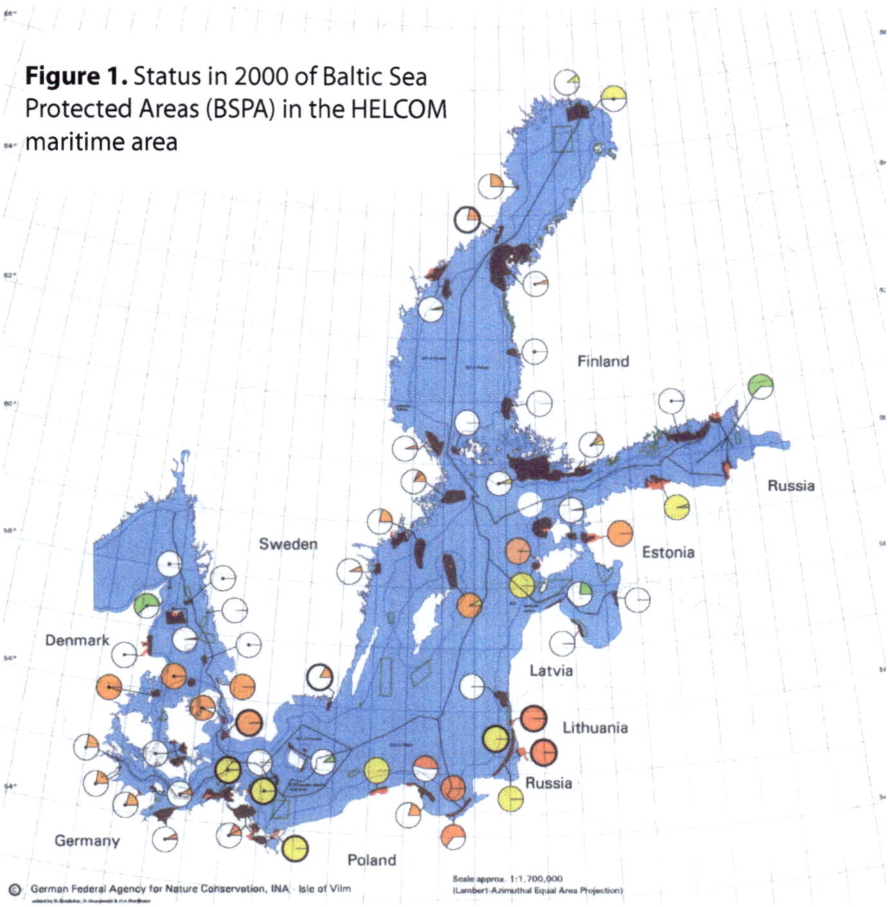

Figure 1. Status in 2000 of Baltic Sea Protected Areas (BSPA) in the HELCOM maritime area

Legend

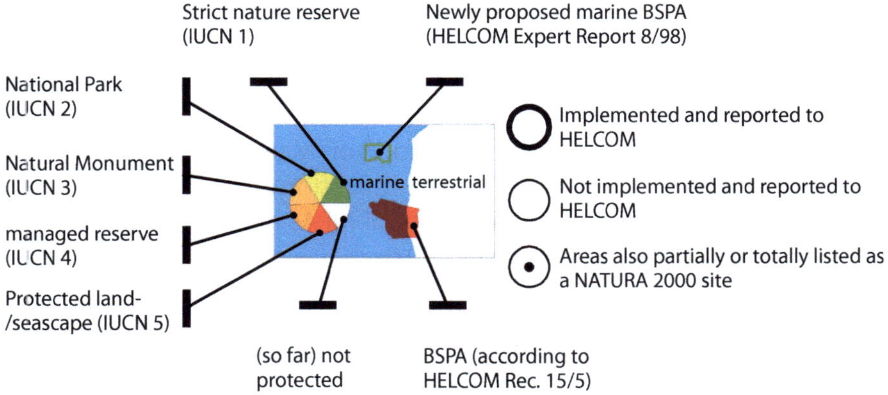

biotopes (von Nordheim and Boedeker 1998) and by various other Red Lists of Baltic species.

There still remains a great challenge for the contracting parties and HELCOM to fully implement the BSPA system and to ensure that by 2010 there will be, together with the NATURA 2000 and the OSPAR sites, an ecologically coherent network of well-managed marine protected areas for the Baltic Sea and the North East Atlantic, as committed to by their Environment Ministers during the joint Ministerial Meeting of the two Commissions in Bremen, Germany 2003 (JMM 2003a) (see also section 3.5 below).

3.4 Convention for the Protection of the Marine Environment of the North-East Atlantic (OSPAR Convention)

The Convention for the Protection of the Marine Environment of the North-East Atlantic (OSPAR, Paris 1992) entered into force in 1998. It combined and updated the 1972 Oslo Convention on dumping waste at sea and the 1974 Paris Convention on land-based sources of marine pollution. It covers the maritime area from the North Pole along Greenland to the Azores and the entire western European coasts and comprises, apart from Territorial Seas and EEZs, about 40% of areas beyond national jurisdiction (High Seas areas; see figure 2). OSPAR's objectives include, to *"protect the maritime area against the adverse effects of human activities..., to conserve marine ecosystems and, when practicable, to restore marine areas which have been adversely affected."* The OSPAR Secretariat in London and the Commission, made up of 15 States and the European Union as parties to the Convention, manage the work under the Convention. Unlike, for instance, the Helsinki Convention, both non-binding recommendations and legally binding decisions can be agreed upon under this Convention (Art. 13).

At a ministerial meeting in 1998 in Sintra, Portugal, OSPAR adopted a new Annex V on the Protection and Conservation of the Ecosystems and Biological Diversity of the Maritime Area, and a respective OSPAR strategy. Consequently in 2003, the Commission adopted OSPAR Recommendation 2003/3 on a Network of Marine Protected Areas, together with the Guidelines for the Identification and Selection of MPAs and the Guidelines for the Management of MPAs in the OSPAR Maritime Area (von Nordheim and Boedeker 2002).

OSPAR network of MPAs

Germany is the lead country in the process of establishing a MPA network in the OSPAR Maritime Area. This task is being carried out concurrently

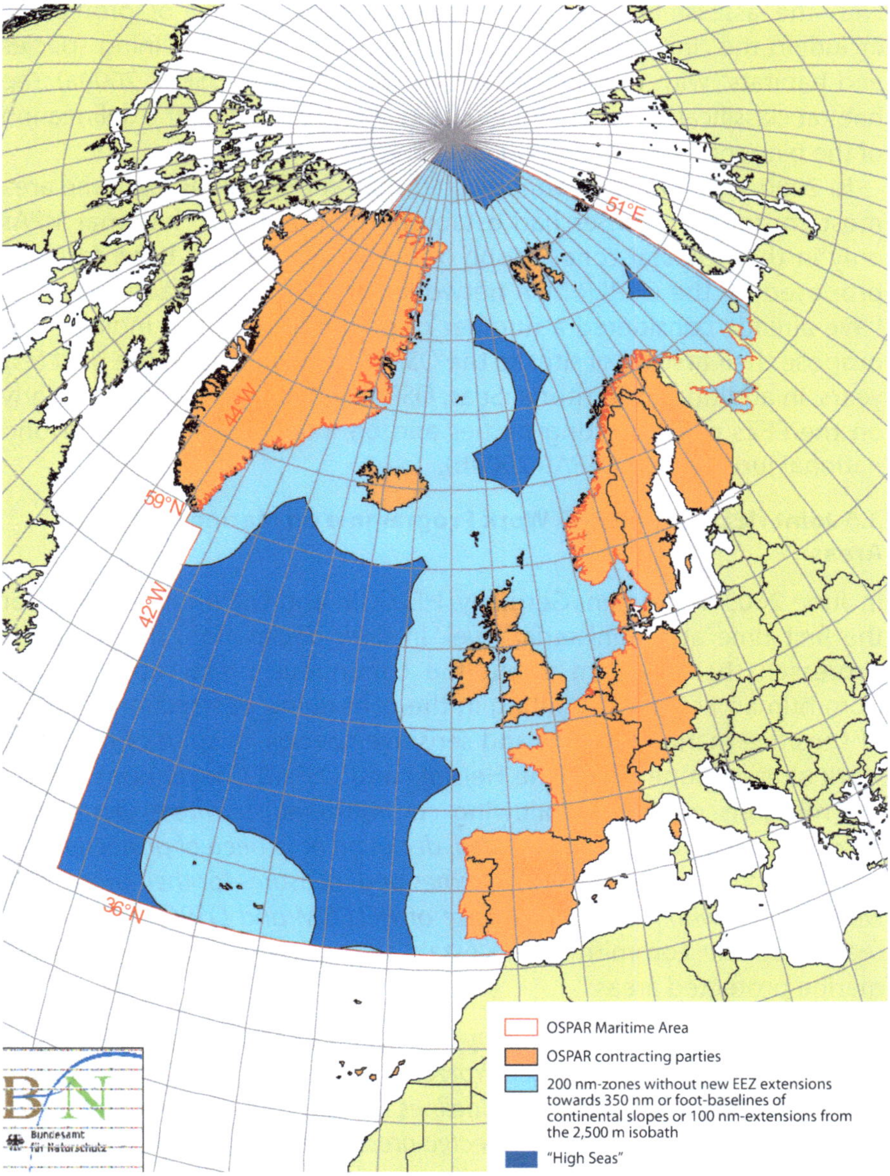

by Wolfgang Dinter, Annette Groß, Federal Agency for Nature Conservation. December 2003

Figure 2. The OSPAR maritime area, covering EEZs and areas beyond national jurisdiction (High Seas)

with the assessment of species and habitats in need of protection, including the "Initial OSPAR List of threatened and/or declining species and habitats" with the Netherlands as lead country (OSPAR 2004a), the habitat classification and mapping (lead country UK), and the delineation of the biogeographic regions of OSPAR, after Dinter (2001).

In 2000, a first evaluation indicated that in the OSPAR maritime area most contracting parties had established a number of national MPAs only in their coastal zones. None of these sites has yet been formally proposed for the OSPAR MPA network, but several contracting parties have already considered a number of potential sites and indicated to propose their first set of sites to the OSPAR network within the next few years. In addition to Territorial Seas, OSPAR will concentrate particularly on the EEZs of contracting parties and on international areas beyond national jurisdiction (OSPAR 2004b).

3.5 Joint HELCOM/OSPAR Work Programme on Marine Protected Areas

In June 2003 in Bremen, Germany, HELCOM and OSPAR met jointly for the first time at a ministerial level. In their declaration, the ministers expressed their concern about the continuous pressures that arise from human activities resulting in the continuing destruction and loss of sensitive marine habitats and sensitive species, in spite of the good work already done under the Helsinki and OSPAR Conventions. Hence the ministers made the following commitment: "*...where species and habitats are identified as threatened, declining or in need of protection, we will take action to develop programmes and measures for their protection, where this is within the competence of HELCOM and OSPAR.*" They also reaffirmed the commitment to establish a network of well-managed marine protected areas:

> *Based on the progress made by HELCOM in establishing a system of coastal and marine Baltic Sea Protected Areas, and OSPAR's agreement to a Recommendation and guidelines for selecting and managing an OSPAR Network of marine protected areas, working with the European Community, we shall have identified the first set of such areas by 2006, and shall then establish what gaps remain and complete by 2010 a joint network of well-managed marine protected areas that, together with the NATURA 2000 network, is ecologically coherent* (JMM 2003b).

In order to concentrate the efforts and facilitate successful results, HELCOM and OSPAR decided to set up a comprehensive joint work programme on the creation of a network of well-managed marine protected areas by 2010 (table 3).

Table 3. Work programme for the establishment of a well-managed ecological coherent network of marine protected areas jointly adopted by the Helsinki Commission and the OSPAR Commission 2003 in Bremen

1 The purpose of the work programme is to ensure that by 2010 there is an ecologically coherent network of well-managed marine protected areas for the maritime areas of both HELCOM and OSPAR ("the network").
2 To these ends, HELCOM and OSPAR will: **a** develop co-ordinated approaches to: i. the compilation and evaluation of proposals for the components of HELCOM and OSPAR networks of marine protected areas, and ii. identifying and addressing any gaps to be filled in order to achieve the network, which reflect the needs for protection of species and habitats identified by HELCOM and OSPAR as threatened, declining or in need of protection; **b** develop and implement a strategy for achieving dialogue with relevant stakeholders on the management and conservation of marine protected areas, using (where possible) existing national and international forums; **c** in order to ensure the ecological coherence of the network, develop common theoretical and practical aspects of what would constitute an ecologically coherent network of marine protected areas; **d** develop, by 2005, a common proposal, taking into account the work done by HELCOM and OSPAR as well as work by the European Community, for a programme aimed at enhancing the protection of species and habitats in European marine waters, in order to produce suggestions for consideration by the European Commission for amendments to the annexes to the Habitats and Birds Directives; **e** consider how Baltic Sea Protected Areas and components of the OSPAR Network of marine protected areas in the waters under the jurisdiction of EU Member States, together with the NATURA 2000 network, can constitute a coherent network of marine protected areas; **f** by 2006, evaluate whether the Baltic Sea Protected Areas and the components of the OSPAR Network of marine protected areas that have been identified by that date are sufficient to constitute the joint network, and take steps to identify and fill any gaps that are identified; **g** by 2010, evaluate whether the aim of establishing the network has been achieved, take steps to fill any shortfalls and to maintain and develop the network thereafter and periodically evaluate whether the aims of the network continue to be met; **h** develop practical guidance for the application of HELCOM and OSPAR management guidelines, including appropriate means to enlist the help of other authorities which are competent for some necessary action, in order to achieve a common standard of good management across the network;

i develop guidance on, and make arrangements for, the assessment of how effectively the management of HELCOM and OSPAR marine protected areas is achieving the aims of protection;

j consider how to take into account other relevant HELCOM and OSPAR initiatives, such as that on the identification and compilation of lists of habitats and species in particular need of protection, and those on marine habitat classification and mapping;

k as appropriate, identify and assist where collaboration with other international forums (such as the Convention on Biological Diversity, and the Berne, Bonn and Ramsar Conventions) may be required, for the implementation and management of HELCOM and OSPAR marine protected areas;

l explore the possibilities for collaboration with the Barcelona Convention and the Bucharest Convention and in the framework of the Arctic Council in this field.

3 To facilitate this joint work, the relevant HELCOM nature conservation working group and the OSPAR intersessional correspondence group on marine protected areas may hold joint meetings, in accordance with arrangements agreed by HELCOM and OSPAR. Where possible, these groups will also work in cooperation with the European Commission and any relevant informal advisory groups that the European Commission establishes.

3.6 The European Network of Protected Areas – NATURA 2000

Two essential EU Directives have been the driving force for the last decade for nature conservation and biodiversity protection in the EU Member States: the Birds Directive and the Habitats Directive (table 1).

In 1992, the European Directive on the conservation of natural habitats and of wild fauna and flora (the Habitats Directive 92/43/EEC of 21 May 1992) entered into force. The principal aim of the Habitats Directive is to protect biodiversity and with this, to contribute to the implementation of the Convention on Biological Diversity in EU-Europe to land and marine environment. In this context, the Habitats Directive ruled on the achievement and continuation of a *favourable conservation status* of natural habitats and species of wild marine fauna and flora of community importance. Moreover, with its enforcement, it became obligatory for Member States to identify sites to be included into a European coherent network of protected areas, i.e., NATURA 2000 (see chapter 2).

NATURA 2000 consists of the Special Protection Areas (SPAs) established under the EU Birds Directive (79/409/EEC of 2 April 1979), and the Special Areas of Conservation (SACs) designated under the Habitats Directive (table 1).

In an official communication on fisheries management and nature conservation in 1999, the European Commission stated clearly that the Birds and Habitats Directives do apply outside Territorial Waters when a Member State has either designated an EEZ or is exercising its sovereignty within the 200-nautical mile zone (Thissen 2002).

The majority of the EU Member States still lack sufficient site proposals in marine areas, particularly in the EEZs (O'Briain and Rizo-Martin 2002). As of May 2004, only Portugal, Denmark, and Sweden have designated SPAs or SACs outside their Territorial Waters. In May 2004, after an amendment to the German Federal Nature Conservation Act, Germany nominated to the EU Commission a complete set of NATURA 2000 sites in its EEZ under the Habitats and Birds Directives, and remains the only EU Member State to have done so.

Some of the following chapters in this book will explain the rationale, and the legal, administrative and scientific steps taken that led to the nomination of ten NATURA 2000 sites in the German EEZ of the North Sea and the Baltic Sea. These sites account for 31.5% of the German EEZ, and together with the sites within the Territorial Sea, even a larger proportion of the total German marine area is nominated, namely 38%.

In conclusion, the EU Nature Directives have proven to be powerful but slow instruments for the establishment of MPAs in European seas. In fact, the Habitats Directive/NATURA 2000 sites should have been already established by 1998. After a period of being regarded as a side issue, the European NATURA 2000 marine areas received a boost in 2003 when the responsible EU administration, Directorate General Environment, began to encourage the marine site nomination process of the EU Member States. This led to the establishment of a Marine Expert Working Group under the EU Habitats Committee, which will, among other tasks, give guidance on the establishment of NATURA 2000 in marine offshore areas, with the aim of completing the marine elements of the NATURA 2000 network by 2008.

4 Conclusions and Outlook

In 1996 the World Conservation Monitoring Centre (WCMC) recorded worldwide about 1.5 million km^2 of marine protected areas, compared with some 11.6 million km^2 protected on land (CBD 2004). Considering that 70% of the world's surface is covered by oceans, this very obvious lack of MPAs needs to be remedied at both regional and global levels.

Although the progress in establishing the BSPA system in the Baltic Sea has been quite slow so far, this integration of marine nature conservation issues into the work of the Helsinki Convention has served as a model for similarly amending other international regional marine conventions on issues of nature conservation, such as OSPAR and the Barcelona Convention.

In the end, it was the legal impetus of the EU Birds and Habitats Directives that caused EU Member States to join forces and come up with site proposals for the NATURA 2000 system in their maritime areas. A joint task force of OSPAR and HELCOM began work in 2003 to develop an ecologically coherent network of well-managed marine protected areas with a target of full implementation by 2010. It will also take into account efforts completed under the EU Habitats and Birds Directives. This gives hope that by 2010 MPA networks will be realised not only in the North Sea and the Baltic Sea, but also in the North-East Atlantic and in parts of the Mediterranean and the Black Sea.

Additionally, the presidency conclusions of the European Council meeting in Göteborg 2001 (European Council 2001) set a target to stop the marine biodiversity decline in the EU by 2010. In order to meet these and other marine nature conservation objectives as outlined in paragraph 2, a number of additional tools and measures, apart from MPAs, will also have to be applied.

Acknowledgements

We would like to thank Jeffrey Ardron (Germany) and Dan Laffoley (UK) for critically reading and commenting on the manuscript.

References

Convention on Biological Diversity (CBD) (2004) Technical Advice on the Establishment and Management of a National System of Marine and Coastal Protected Areas – Technical Series 13

Convention on Biological Diversity (CBD) (2005) CBD website, http://www.biodiv.org/programmes/areas/marine/default.asp, last accessed June 2005

De Fontaubert, AC, Downes DR, Agardy TS (1996) Biodiversity in the Seas: Implementing the Convention on Biological Diversity in Marine and Coastal Habitats. IUCN Gland and Cambridge

Dinter WP (2001) Biogeography of the OSPAR Maritime Area. Federal Agency for Nature Conservation, Bonn

European Environment Agency (EEA) (2001) Marine and Coastal Environment – Annual Topic Update 2000. Luxembourg, Office for Official Publications of the European Community

European Council (2001) Presidency Conclusions. Meeting of the Council in Göteborg, Sweden, 15 and 16 June 2001

Gjerde KM (2002) International Legislation in Support of Implementing NATURA 2000 in the Marine Environment. In: Boedeker D, von Nordheim H (eds) Application of NATURA 2000 in the Marine Environment. BfN-Skripten 56:73–77

Hägerhäll B, Skov H (1998) Proposal for offshore Baltic Sea Protected Areas (BSPAs). Unpublished expert report to HELCOM EC-NATURE

HELCOM (1996) Coastal and Marine Protected Areas in the Baltic Sea Region. - BSEP 63

HELCOM (2002) Fourth Periodic Assessment of the State of the Marine Environment of the Baltic Sea Area, 1994–1998. Background Document. BSEP 82B

HELCOM (2003) The Baltic Marine Environment 1999–2002. BSEP 87

International Council for the Exploration of the Sea (ICES) (2003) Environmental Status of the European Seas. A Quality Status Report prepared for the German Federal Ministry for the Environment, Nature Conservation and Nuclear Safety

International Marine Organization (IMO) (2005), http://www.imo.org/home.asp, last accessed June 2005

Joint Ministerial Meeting (JMM)(2003a) Statement on the ecosystem approach to the management of human activities. Joint Ministerial Meeting of the Helsinki and the OSPAR Commissions, Bremen, Germany, 25 to 26 June 2003. Record of the Meeting, Annex 5

Joint Ministerial Meeting (JMM) (2003b) Declaration of the first Joint Ministerial Meeting of the Helsinki and OSPAR Commissions. Joint Ministerial Meeting of the Helsinki and the OSPAR Commissions, Bremen, Germany, 25 to 26 June 2003. Record of the Meeting, Annex 8 (http://www.helcom.fi./ministerial_declarations/en_GB/ospar-declaration/)

Kelleher G (ed) (1999) Guideline for Marine Protected Areas. IUCN, The World Conservation Union – Best Practice Protected Area Guidelines Series 3

von Nordheim H, Boedeker D (1998) Red list of marine and coastal biotopes and biotope complexes of the Baltic Sea, Belt Sea and Kattegat. Including a comprehensive description and classification system for all Baltic marine and coastal biotopes. BSEP 75

von Nordheim H, Boedeker D (2002) Other International Initiatives in the Management of Marine Protected Areas (OSPAR and HELCOM). In: Boedeker D, von Nordheim H (eds) Application of NATURA 2000 in the Marine Environment. BfN-Skripten 56:80–89

O'Briain, M, Rizo-Martin J (2002) The state of implementation of nature directives with particular reference to the marine environment – Introduction and overview of the situation. In: Boedeker D, von Nordheim H (eds) Application of NATURA 2000 in the Marine Environment. BfN-Skripten 56:49–51

OSPAR (2000) Quality Status Report 2000. OSPAR Commission, London

OSPAR (2004a) 2004 Initial OSPAR List of Threatened and/or Declining Species and Habitats. OSPAR Agreement 2004 – Report of the OSPAR Commission meeting 2004, Reykjavik, Iceland. Annex 5

OSPAR (2004b) Meeting of the Working Group on Marine Protected Areas Species and Habitats (MASH). Summary Record. OSPAR Commission, London

Salm RV, Clark JR (2000) Marine and Coastal Protected Areas. A Guide for Planners and Managers. IUCN, Gland, Switzerland

SRU – The German Advisory Council on the Environment (2004) Marine Environment Protection for the North and Baltic Seas. Special Report. Nomos, Baden-Baden

Thissen J (2002) NATURA 2000 and the Protection of Marine Biodiversity: Current Status and further Prospects. In: Boedeker D, von Nordheim H (eds) Application of NATURA 2000 in the Marine Environment. BfN-Skripten 56:52–53

Chapter 2

Marine Protected Areas in the EEZ in light of international and European Community law – Legal basis and aspects of implementation

Detlef Czybulka and Thomas Bosecke

University of Rostock, Faculty of Law, Rostock

Abstract

At the international level, Articles 192, 194, and 56(1)(b)(iii) of the United Nations Convention on the Law of the Sea (UNCLOS) oblige the coastal States to protect the marine environment in their own Exclusive Economic Zones (EEZs). The measures required under international law also include the establishment of Marine Protected Areas (MPAs). Regulations in MPAs must be based on the sovereign rights and jurisdictions given to the coastal States by UNCLOS. Admissible restrictions concern most forms of economic uses such as all kinds of installations, the exploration and exploitation of the living and non-living resources in the water, seabed and subsoil. Marine scientific research is also covered by such restrictions, but not navigation, overflight and military use. Specifications of this rather general obligation derive from regional or global international environmental law.

The habitat protection directives under European Community legislation are legally enforceable and sanctions-implying obligations to carry out site protection. In the framework of the sovereign rights and jurisdictions that UNCLOS assigns to the Member States, the latter are obliged by the directives to establish even in their EEZs the coherent ecological network of protected areas known as NATURA 2000. The selection of the sites follows exclusively technical and scientific criteria. The protection system substantially follows Articles 6(2), 6(3), 6(4) and 7 of the Habitats Directive (HD). In accordance with these provisions, plans and projects which may adversely affect the site shall only be agreed to if, in light of the precautionary principle, no reasonable scientific doubt remains as to the absence of such effects. Possible exceptions must strictly follow the provisions under Articles 6(3) and 6(4) of the HD. Protection does not only have to be guaranteed at the time when an authorisation or licence is granted, but permanently.

In Germany, site protection in the EEZ is implemented through Article 38 of the Federal Nature Conservation Act (BNatSchG) and in the form of relevant statutory ordinances on protected areas. Although Article 38(1)(3) BNatSchG refers to the EU legislation, Germany is responsible for the regulation of fisheries within the MPAs. This is because Member States are responsible for issuing site-related protection provisions – even if these have side effects on fishery – when fulfilling their protection obligation under Article 6 HD and Article 4 of the Birds Directive (BD). Based on Article 6 EC Treaty, the Community can also take measures under the Common Fisheries Policy to support the Member States in their efforts to protect species and habitats in their NATURA 2000 sites (e.g., protection of the *Darwin Mounds*). Thus, the Council of the EU can adopt measures with nature conservation effects, but it can never supersede the Member State in their own responsibility. The restrictions under Articles 38(1)(4) and 38(1)(5) BNatSchG rule out the possibility of abstract and general prohibitions of projects mentioned in that Article (prospecting and extraction of mineral resources, windmills, etc.) but not the duty to carry out an impact assessment. Statutory ordinances with regard to protected areas in the EEZ are to be implemented by use of management plans under the responsibility of the Federal Agency for Nature Conservation (BfN).

1 The basis and framing conditions of MPAs in international law

1.1 The United Nations Convention on the Law of the Sea

1.1.1 Mission and obligation to protect the marine environment

The protection of the marine environment through international laws has evolved since 1945. In the beginning, environmental treaties regulated individual aspects of the pollution of the marine environment and the protection of living marine resources (Beyerlein 2000, para. 220ff). Today, Part XII of the United Nations Convention on the Law of the Sea (UNCLOS)[1], deemed as the constitution of the seas, provides for the basic

[1] Entered into force 16 November 1984; Federal Law Gazette of 10 December 1982, 21 ILM 1982, p. 1261, [Bundesgesetzblatt] 1994 II p. 1799.

rights and duties of the coastal States and of the international community of States in the ocean, and it includes a comprehensive mission to protect the marine environment. Article 192 UNCLOS expressively provides for the States to protect and preserve the marine environment. This is a genuine legal obligation having even *erga omnes*[2] effect (cf. Proelß 2004, p. 77), however, its scope has been controversial. Based on Article 194 paragraph 5 UNCLOS, this mission includes all measures necessary to protect and preserve rare or fragile ecosystems as well as the habitat of depleted, threatened or endangered species and other forms of marine life. Just as with any other constitution, UNCLOS needs to be concretised and implemented through national laws, regional protection systems (cf. Articles 197 to 201 UNCLOS), or by empowering an international organisation.

UNCLOS assigns the coastal States certain sovereign rights and jurisdiction – which are basically of the same kind – in the EEZ. These are either exclusive or primary-interest rights and jurisdiction (e.g., exploiting the living resources) (Jarass 2002, p. 22; Stoll 1999, p. 668). The sovereign rights and jurisdiction (and duties) of the coastal States in the EEZ are listed under Article 56 UNCLOS and are described in more detail for artificial islands, installations and structures under Article 60.[3] Article 61 formulates the rights and duties with regard to the conservation and management of the living resources of the EEZ in a very elaborate manner. The sovereign rights and jurisdictions provided in Article 56(1)(a) and (b) need to be claimed by the coastal State in order to be exercised. In practice, this has been done by the proclamation of an EEZ by Germany and by most other coastal States. The United Kingdom has not done this until today (April 2005). In contrast to that, the sovereign rights under the continental shelf regime (Part VI Art. 77 to 81 UNCLOS) do not depend on occupation, effective or national, or on any expressed proclamation (Art. 77 para. 3). The continental shelf of a coastal State comprises (only) of the seabed and the subsoil of the submarine area, and not the body of water, up to a distance of 200 nautical miles from the baselines, and in some cases even up to 350 nautical miles (Art. 76 para. 10). On the continental margin, coastal States have *inter alia* sovereign rights to explore and exploit the mineral and other non-living resources of the seabed and subsoil, including living organisms belonging to sedentary species (Art. 77 para. 4 UNCLOS). Article 56 paragraph 3 links the EEZ and continental

---

[2] *Erga omnes*: literally 'between all', in this context: affecting all states and their nationals.

[3] Article 60 applies *mutatis mutandis* to artificial islands, installations and structures on the continental shelf (Art. 80 UNCLOS).

shelf regime by stating that the rights set out in Article 56 UNCLOS shall be exercised in accordance with Part VI. In the following, the continental shelf regime will not be covered in detail, although some relevant aspects on species protection are mentioned.

The tasks of protection mentioned under Part XII of UNCLOS are implemented by Article 56(1)(b)(iii) UNCLOS into the legal order of the EEZ and transferred into the coastal States' competence. The rights and jurisdiction assigned to the coastal States are particularly important for the establishment of MPAs (see 1.1.2 below for further detail). Outside the scope of special rights and jurisdiction the legal order of the EEZ essentially follows the legal regime of the high seas (Article 87ff UNCLOS) in accordance with Article 58 UNCLOS. This applies in particular to the freedoms of navigation and over-flight and (in parts) to the freedom of the laying of submarine cables and pipelines and other internationally lawful uses of the sea related to these freedoms.[4]

However, it is not at the discretion of the coastal States to claim the rights and yet not assume the tasks and duties resulting from UNCLOS for the protection of the marine environment. Those who make use of the rights of economic exploration and exploitation are obliged to protect, as a countermove, the marine environment in accordance with UNCLOS.[5] This is also clarified by Article 193 UNCLOS, which obliges the States to link simultaneously the exploitation of their natural resources to their very protection.

Possibly there are some or many strategies and measures for successfully implementing the protection of ecosystems stipulated under Article 194(5) UNCLOS. Czybulka and Kersandt (2000, p. 7) think that the inclusion of three-dimensional area strategies in this context is essential: in some cases, it might be sufficient to protect the seabed and/or deep layers of the body of water. Such strategies also call for the creation of biologically needed disturbance-free No-Take-Zones or other qualified areas for retreat and reproduction. Thus the basic obligation to protect and preserve the marine environment (Art. 192) becomes more concrete and converts to an obligation of habitat protection (Art. 194(5)) by establishing MPAs when rare and fragile ecosystems can only thus be protected in accordance with Article 194(5) UNCLOS (see Wolfrum 2001, p. 430). There are no more disputes in literature as to whether it is basically admissible to assign MPAs in the EEZ based on international law.[6] In

4 For the continental shelf, compare Article 79.

5 Czybulka 2001a, p. 369; Long and Grehan 2002; Kersandt 2002, p. 121; Nebelsiek 2002, p. 10; Gellermann 2004, p. 76.

6 Czybulka 1999, pp. 563f; Jarass 2002, pp. 29ff; Proelß 2004, pp. 91ff.

practice, coastal States have also been doing just this (see Janssen 2002, pp. 76ff for examples).

The older existing literature on international law contested the coastal States' jurisdiction to establish MPAs in the EEZ. One basic misunderstanding was to confuse the establishment of MPAs with the assertion of territorial ('aquitorial') claims. There is no link to territorial claims, even if geographic references are necessary for marking the protected areas. The International Maritime Organisation's (IMO) exclusive regulatory competence with regard to shipping (Wolfrum 2002, p. 7; similar in Lagoni 2002, p. 128), served as another starting point for conflicting opinions. MPAs need not necessarily be legally or technically strict nature reserves (Czybulka 2001a, pp. 373f). In many cases, it would not make sense to exclude navigation in a MPA. Rather, MPAs in the EEZ are restrictions of use within the jurisdiction assigned to the coastal State (similar in Lagoni 2002, pp. 123f). The IMO's exclusive competencies only correspond to navigation-related rules. Such rules will only be necessary in the EEZ in exceptional cases. Although shipping activities cannot be legally regulated by the coastal States, the protection of marine areas can be achieved by regulating human activities using the powers already granted to the coastal States in the EEZ for protecting and conserving the marine environment.

Such restrictions may, of course, also be "bundled" and applied to certain marine areas in the EEZ. Marine protected areas are thus established (Czybulka 2001a, p. 374).

1.1.2 Regulatory power

As mentioned above, it is necessary to regulate the different types of human activities that could be harmful to the designated MPAs. Decisive in this respect is the question of whether it is admissible in terms of international law to create area-related limitations. For the EEZ, rights and jurisdiction for conserving and managing the natural resources are derived from Article 56, Article 60 (concerning installations and structures), for fishing from Article 61ff, in particular from Article 61(2), and for the continental shelf from Articles 77 and 80 UNCLOS. Subject to certain restrictions, the coastal State may autonomously regulate the exercise of these assigned rights and jurisdiction.[7] Coastal States may therefore regulate the different types of use and human activities in the form of a regulation which applies to all states and nationals (*erga omnes*).

[7] Klinski 2001, pp. 9ff; Nebelsieck 2002, pp. 8ff; Wolf and Heugel 2001, pp. 8ff.

They are not limited to controlling the behaviour of their nationals; the flag States have no special rights beyond Article 73(4).

The question is which additional activities the coastal States may regulate in the EEZ. Article 58(1), in conjunction with Article 87 UNCLOS, shows that it is not possible for the coastal State alone to regulate navigation and over-flight in a way admissible in international law. It is possible for the coastal State to submit an application for navigation rules at the IMO in accordance with Article 211(6)(c) UNCLOS concerning the ecological conditions of the area. The IMO may then allow the coastal State to apply (even) stricter rules to its EEZ with regard to the pollution of the area from vessels (for more details and in relation to "Particularly Sensitive Areas", see Proelß 2004, pp. 89ff).

The coastal State may, in accordance with Articles 87 and 79(4) UNCLOS, establish (legal) conditions for the laying of submarine cables or pipelines entering its territory or territorial sea, or it may establish its jurisdiction over cables and pipelines constructed or used in connection with the exploration of its continental shelf or the exploitation of its resources or the operations of artificial islands, installations and structures under its jurisdiction. The laying of transit *pipelines* requires the consent of the coastal State in accordance with Article 79(2)(3) UNCLOS and thus, it can also be regulated within a MPA. However, it will most probably not be possible to provide for the abstract regulation (prohibition) of the laying of transit *cables* since such cables are not covered by the requirement of consent (subject to the consent) in the text of Article 79(2)(3) UNCLOS (controversial, Lagoni 2002, p. 124; see also Wolf 2004, p. 71). However, it should be possible for the coastal State to influence the type of cable laid, according to Article 79(2) UNCLOS, and the location (the delineation of the course) where the cables will be laid.

According to Article 56, the coastal State has sovereign rights for resource-related marine scientific research (i.e., exploration and exploitation). Coastal States may, in accordance with Article 246(5) UNCLOS, withhold their consent to marine scientific research activities which may adversely affect the sovereign rights for the exploration and exploitation of the natural resources. Such research may thus also be regulated in a MPA. In principle, the coastal State must tolerate marine scientific research unrelated to living resources (pure research) in its EEZ and on the continental shelf with one exception: restrictions are possible if the existing circumstances are not "normal circumstances" within the meaning of Article 246(3) UNCLOS. Normal circumstances may be ruled out for MPAs if it can be proven that they are actually worthy of protection and require special protection, and if the intended research activities would adversely affect the area (Stoll 2004, p. 50). The result is that resources-unrelated marine scientific research (pure research) may also be regulated significantly in MPAs.

1.2 International environmental law

Modern international environmental law constitutes the second pillar of the mission and obligation to establish MPAs. At the global level, the Convention on Biological Diversity (CBD) of 5 June 1992 (Bundesgesetzblatt 1993, II, p. 1741) is of great importance. According to Article 8(a) CBD the Contracting Parties are called upon to establish, as far as possible and as appropriate, a system of protected areas or areas where special measures need to be taken to conserve biological diversity. Upon ratification of the Convention, the Federal Republic of Germany assumed the obligation to implement the provisions contained therein.[8] The CBD neither modifies nor extends the regulations of UNCLOS.

Specifications of the CBD for the North Sea and the Baltic Sea derive from the Helsinki Convention on the Protection of the Marine Environment of the Baltic Sea Area (HC) (of 9 September 1992, Bundesgesetzblatt 1994 II, p. 1355) and from the Oslo Paris Convention for the Protection of the Marine Environment of the North-East Atlantic (OSPAR) (of 22 September 1992, Bundesgesetzblatt 1994 II, p. 1355). For more details on international conventions, see chapter 1.

2 Standards in European Community legislation

2.1 The coherent ecological network NATURA 2000

Recognising that the decline of biological diversity can only be combated through coordinated Europe-wide protection measures, the Council of the European Community decided to set up a coherent European ecological network of special areas of conservation known as NATURA 2000 (see Art. 3(1) sentence 1 of the Habitats Directive). Through the designation of protected coherent areas composed of sites hosting the natural habitat types listed in Annex I and habitats of the species listed in Annex II, the network shall serve the establishment of disturbance-free areas, the genetic exchange of certain wild species of fauna and flora, and thus the conservation of the European natural heritage.[9] The

[8] Czybulka 2001, pp. 24ff; Wolff 2004, p. 177; Jarass 2002, p. 40.

[9] Gellermann 1996, p. 548; Ssymank et al. 1998, pp. 11f.

system of NATURA 2000 areas comprises of two area types: Special Areas of Conservation (SACs) under the Habitats Directive (HD)[10], and Special Protection Areas (SPAs) under the Birds Directive (BD).[11] SACs are selected in detailed procedures from Sites of Community Importance (SCIs) proposed by the Member States (pSCIs). The Habitats Directive implements the Berne Convention (of 19 September 1979, Bundesgesetzblatt II 1984, p. 618) at the European Community level, and with the protection of habitats, it provides an extension to the legal tools of the said Convention (Wagner 1990, p. 396; Bosecke 2005, p. 334). The Birds Directive, on the other hand, is the legal adaptation of the Ramsar Convention at the Community law level (Czybulka 1996, p. 48). The NATURA 2000 Network is generally understood to be the European implementation of the task contained in Article 8(a) CBD (Czybulka 2003, p. 62) which requires a "system" of protected areas to be established.

With respect to the protection aim under Article 2 Habitats Directive for all areas incorporated in the NATURA 2000 Network, the Member States must take both statutory (legal) and administrative measures to maintain or restore the favourable conservation status of the natural habitats and species of wild flora and fauna. Member States have to classify the most suitable territories (in number and size) as SPAs for the conservation of bird species listed in Annex I,[12] and of migratory bird species not listed in Annex I[13] with regard to their geographical distribution on sea and land, and to their breeding, moulting and wintering areas and staging posts. For site selection under the Habitat Directive (SAC), the relevant criteria are the priority and non-priority species listed in Annex II and the natural habitat types listed in Annex I of that Directive. From the expert point of view, the Annexes fail to be complete with regard to the marine area; Annex II, for example, does not include any marine water plants.

2.2 Does European nature conservation and environmental legislation apply to the EEZ?

Recent discussions and jurisprudence have clarified that the Birds Directive and the Habitats Directive are applicable to the EEZ (and the continental shelf) of the Member States.[14]

---

[10] Habitats Directive: 92/43/EEC, OJ EC L 206, p. 7.

[11] Birds Directive: 79/709/EEC, OJ EC L103, p. 1.

[12] Under Article 4(1) sentence 4 of the Birds Directive.

[13] Under Article 4(2) Birds Directive.

[14] London High Court, LO 1336/1999 NuR 01, pp. 19ff; European Commission, evidence ibidem; Czybulka 2001, pp. 19ff; Kersandt 2002, p. 124; Jarass 2002, pp. 41ff; Long and Grehan 2002, pp. 248f; Owen 2004, p. 61.

The HD mentions, among others, aquatic areas in Article 1(b) and aquatic species, which range over wide areas, in Article 4. At least the following natural habitat types in need of protection as listed in Annex I of the HD can be found in the German EEZ: sandbanks (Annex I HD, NATURA-2000-Code No. 1110), and reefs (Code No. 1170). In addition, some species listed in Annex II are known to regularly occur in the German EEZ; these include harbour porpoise (*Phocoena phocoena*), grey seal (*Halichoerus grypus*), common seal (*Phoca vitulina*), river lamprey (*Lampetra fluviatilis*), sea lamprey (*Petromyzon marinus*), and twaite shads (*Alosa fallax*). These species and habitats would therefore not be effectively protected if the scope of application were limited to the Member States' Territorial Waters, excluding the EEZ (Nordberg 2004, p. 116; for the Directive on Environmental Impact Assessment: Jarass 2002, p. 49). The same applies to the bird species listed in Annex I and the migratory species according to Article 4(2) that are protected under the BD.

2.3 The obligation to protect habitats

2.3.1 The site selection

The Birds Directive and the Habitats Directive specify the protection obligations more explicitly than the international environmental law. Protection starts with certain natural conditions, which at the EU level, have been assessed as sites in need of protection in accordance with Article 4(1) BD ("most suitable territories") and in accordance with Annex III of the HD (habitat types listed in Annex I and species listed in Annex II). The directives are legally enforceable and imply sanctions (Czybulka and Kersandt 2000, pp. 26f).

When classifying the "most suitable territories" under Article 4(1) BD, Member States have only a limited scope of selection. In those cases where a territory or aquatic area has been classified as the most suitable in purely ornithological terms with regard to the protection purpose of the BD,[15] the classification must take place. Economic or political reasons are not allowed.[16] In principle, the same applies to the designation of sites under the HD, even if the selection procedure and the criteria to be applied are much more complex. Sites are exclusively selected in line with the technical criteria set out in Annex III.[17] In summary: when

[15] For example, in accordance with the IBA lists drawn up by expert associations, cf. ECJ, Case C-374/98- ZUR 01, p. 76; Iven 1998, pp. 529f.

[16] ECJ, Case C-3/96- NuR 98, p. 539; Iven 1998, p. 529; Spannowsky 2000, p. 43.

[17] Rödiger-Vorwerk 1998, p. 36; Ssymank et al. 1998, p. 23; see also chapter 4.

selecting sites, Member States are not allowed to consider issues other than those of nature conservation, for example, those of an economic or structural nature (ECJ, Case C-371/98- DVBl. 00, p. 1842). The same applies to the final act of the official designation of the site according to German national legislation (Louis 2000, §19 para. 11). Such economic or structural considerations may only be undertaken within the exceptions provided under Article 6(4) HD (cf. Article 34(3) and Article 35 of the Federal Nature Conservation Act (BNatSchG) (see section 2.2.3 above).

The material requirements with regard to site selection contained in the HD and BD are essentially of the same kind. The significant differences between these two Directives and the German national legislation for nature conservation areas other than NATURA 2000 sites, concern the selection criteria and the classification or designation duties. Whereas under the HD only scientific criteria apply for site selection and Member States have the duty to establish protected areas for Annex I habitats and Annex II species, conventional German national law allows a wider scope of criteria, including for example economic and political criteria, and does not oblige the establishment of protected areas on a national basis. Additionally, maintenance of the network protection in EC Law is mandatory, although there are still questions to be answered with regard to the marine area.

2.3.2 The protection level to be guaranteed

The protection system in SACs and classified and protected SPAs (for such SPAs see Art. 7 HD) follows the provisions under Article 6(2) to (4) HD. Under this article, the Habitats Directive does not provide protection to sites that should have been classified as SACs but have not yet been so classified. The Birds Directive, however, under Article 4(4) does provide protection for SPAs not yet classified as such.[18] Article 6(2) HD is the central link to the protection level to be guaranteed. Under this provision, the Member States shall take appropriate steps to avoid in the SACs the deterioration of natural habitats and the habitats of species, as well as significant disturbances of the species for which the areas have been designated, independent of the fact whether such effects were caused within the area or were brought in from outside. Thus, there is a general prohibition of deterioration for which surveillance is to be undertaken (Art. 11 HD).

[18] (ECJ, Case C-374/98- ZUR 01, pp. 76f; Federal Administrative Court, Reference 4 VR 13.00, ZUR 02, p. 226).

Article 6(3) and (4) HD provide for special provisions with regard to *plans* and *projects*. The Habitats Directive does not define plans and projects. The European Court of Justice (ECJ) follows the wider concept of "project" given in the Council Directive 85/337/EEC.[19] Accordingly, mechanical cockle fishing in the Netherlands' Wadden Sea for which new licences have to be applied annually has been assessed as a *project* (ECJ ibid., para. 25). In light of this opinion, projects are not only such classical measures like constructions of residential buildings, infrastructure or industrial structures, but they are also parts of agriculture, forestry or fishing use. Impact assessments with regard to the implications for the site's conservation objectives must be carried out if the plan or project, individually or in combination with other plans and projects, is likely to have a significant effect on the site. The plan or project may be authorised only after the national authorities have ascertained that a plan or project will not adversely affect the integrity of the site concerned. This is only the case where, given the precautionary principle, no reasonable scientific doubt remains as to the absence of such effects (ECJ ibid., para. 58f). Otherwise, plans and projects may not be authorised unless the exceptions provided under the first paragraph of Article 6(4) (for sites not hosting priority habits or priority species) and under the second paragraph of Article 6(4) (for sites hosting priority habits or priority species) apply. These exceptions are: "imperative reasons of overriding public interests", the absence of alternative solutions, and the maintenance of the overall coherence of NATURA 2000 even if a plan or project will be carried out (Article 6(4) HD). The Commission shall be informed of the compensatory measures adopted. Where the site concerned hosts a priority natural habitat type and/or a priority species, and where the assessment of the implications for the site is negative, the only considerations which may be taken into account for carrying out the project are those relating to human health, public safety, or to beneficial consequences of primary importance for the environment, or to other imperative reasons of overriding public interest. In this case, the Commission must deliver its opinion prior to a possible authorisation of the plan or project.

The admissibility of all types of exploitation that are neither plans nor projects is exclusively assessed under both Article 6(2) HD and the specific national provisions adopted to this end. Article 6(2) obliges the Member States to take appropriate steps to avoid the deterioration of natural habitats and the habitats of species in the SACs. Human activities

[19] cf. Judgement of 7 September 2004, Case C-127/02, para. 24.

that could bring about deteriorations of natural habitats and the habitats of species as well as significant disturbances are not allowed. This requires continuous surveillance (cf. Art. 11). Article 6(2) HD aims at maintaining (at least) the conservation status at the date of the site's designation and intends to avoid the creation of such adverse implications for the site which add up to the ones already existing at that time.[20] In this respect, the assumption prevails that legally existing activities or exploitations do not negatively modify the conservation status of the site. Such an assumption, however, does not apply to new forms of exploitation and is refuted for existing ones if the protected assets are damaged largely due to intensified exploitation (ECJ, Case C-117/00). The same will have to apply to existing exploitations which have not been intensified or modified but which will permanently lead to deteriorations or considerable disturbances thus aggravating the existing adverse effects in type and quantity at the date of identification.

Recently, it has been argued that Article 6(2) HD provides for the substantive protection level of protected sites with regard to all effects, including those caused by plans and projects, and that it establishes a permanent obligation for its conservation.[21] The consequence of this is that both exploitations (those that are not plans or projects), and plans and projects that have been drawn up and implemented on the basis of valid authorisations, shall be submitted to constant control as to whether they negatively modify the conservation status. Such opinion is to be preferred in the interest of the protection of habitats. Only by way of a permanent obligation, including a control function, will it be possible to permanently maintain the coherence of the NATURA 2000 network and thus to ensure the conservation of biological diversity and the protection of the European natural heritage.

The judgement of the ECJ of 7 September 2004 on Case C-127/02 is not opposed to this opinion: once the assessment has been carried out, a "concomitant application of the rule of general protection laid down in Article 6(2)" becomes "superfluous" (ECJ ibid., para. 35). However, this statement presupposes that the authorisation of the project necessarily means that it is not considered likely to give rise to deteriorations (ECJ ibid., para. 36). Hence, the general level of protection and the conservation of habitats and of species remain untouched. If deteriorations occur, they shall be regulated by retroactively (not concomitantly) applying

---

[20] European Commission 2000, pp. 26f; Gellermann 2001, pp. 72f.

[21] AG Kokott, Opinion delivered on 29 January 2004 on Case C-127/02; Federal Administrative Court, Reference 4 C 2.03, p. 18 of the document.

the measures ("steps") of Article 6(2) HD (similar: ECJ ibid., para. 37). This involves far-reaching consequences for valid authorisations (permits or licences) that originally allowed projects or plans in protected areas. Thus, the Member States would have the opportunity to make sure that it would be possible to modify and possibly reverse administrative decisions in order to be able to interfere in cases where Article 6(2) HD has been violated. With regard to Germany, it should be taken into consideration by the administration whether more limited or revocable authorisations will be required for corresponding activities than before. This is because German administrative law grants a strong position to the holder of a once-legal or even illegal authorisation (permit or licence) if that authorisation is valid (entered into effect).

3 Site protection in Germany – implementation of NATURA 2000

3.1 Basic principles

In the Federal Republic of Germany, protected areas for the purpose of nature conservation are as a rule designated by means of legal acts (ordinances). Since such designations have legal effects on third persons, they require a legal basis on the grounds of the constitutional principle of the rule of the law (Article 20(3) of the German Constitution (GG)). Due to the federal structure of Germany, the federal States (Länder) are responsible for the creation of the necessary legal basis in their respective Nature Conservation Acts. The German Federal Nature Conservation Act (BNatSchG) is a piece of framework legislation under Article 75(3) of the German Constitution (GG). It contains some principles and standards for instructions to the federal States and specifies how the Länder have to use their *legislative* competence in their territories and in the adjacent Internal Waters and Territorial Sea. Under Article 30 GG, the Länder have as a rule the full *executive* competencies.

In 2002, Article 38 was introduced into the Federal Nature Conservation Act and thus legal authorisation was established with regard to the EEZ. Hence, the Federal Ministry for the Environment, Nature Conservation and Nuclear Safety is responsible for the establishment of protected areas in the EEZ and on the continental shelf in accordance with some provisions of nature conservation legislation as it applies to the German territory. Under Article 38(2) BNatSchG, the Federal Agency for Nature Conservation (BfN) shall perform the tasks resulting from the establishment and protection of the European Network NATURA 2000 in the EEZ with two relevant exceptions. The exceptions refer firstly, to the implementation of the assessment of plans and projects under the Habitats Directive, and secondly, to the formal designation of areas as "protected parts of nature and landscape". The second exception has to be done by the Federal Ministry for the Environment, Nature Conservation and Nuclear Safety by way of an ordinance (*Rechtsverordnung*). At the same time, it can be derived from this provision that the legislator intended the protection of marine areas to be primarily guaranteed through the adoption of official statutory acts and not through other administrative or contractual measures which are also mentioned under Article 6(1) HD (Gellermann 2004a, pp. 11, 15).

Article 38(1) BNatSchG declares Articles 33 and 34 BNatSchG (Protected Areas, and Impact Assessment, Inadmissible Projects and Exemptions, respectively) applicable to the protection of marine areas in the EEZ.

Thus, the prerequisites established for the most important legal standards of the BNatSchG are also to be applied to the EEZ in order to implement the EC directives on habitat protection. The wording of Article 38(1) and (2) BNatSchG exclusively refers to the implementation of the European Directives on habitat protection. Habitat types which are to be protected according to OSPAR/HELCOM, but not by the EC-Habitats Directive do not fall under the wording of Article 38 BNatSchG (cf. Gassner 2003, sect. 38 para. 8). However, it is not likely that the legislator had such intention because the OSPAR and HELCOM Conventions, and the on-going process of establishing a network of MPAs in their respective areas are firmly supported by the European Community and by Germany. It would be unfair to sign international documents in order to establish a network of marine MPAs on the one hand, and hamper their designation by national law on the other. Hence, a different interpretation or even an amendment of the Act is needed.

Article 33 BNatSchG essentially governs the designation of protected areas, thus serving the implementation of Article 4(1) BD and Article 4(1) HD. Article 33(5) BNatSchG treats aspects of preliminary protection. Article 33(5) BNatSchG and the respective special legal acts of conservation (ordinances) shall implement the general prohibition of deterioration resulting from Article 6(2) HD. Article 34 BNatSchG delivers the special protection and assessment schemes for *projects*, mainly in accordance with Article 6(3) and 6(4) HD. Article 38 BNatSchG does not explicitly refer to Article 35 BNatSchG, which delivers the implementation of *plans*, even though both projects and plans are mentioned in Article 6(3) and 6(4) HD.

This gap might however be considered to be covered by the reference in Article 33(3) BNatSchG to Article 6 HD (Gellermann 2004a, p. 80).

3.2 Contents and provisions in the legal act of conservation

Under Article 22(2) BNatSchG it is necessary, with regard to the effective designation as a protected area (in the format of an ordinance), to determine the object to be protected, the protection purpose, the orders and prohibitions to be complied with to achieve the protection purpose, and, where necessary, the respective management, development and restoration measures. The object to be protected is the detailed description of the area to be protected and the area boundary definitions (Schumacher and Fischer-Hüftle 2003, sect. 22 para. 19). The description of the protection purpose is important to justify that protection is needed and that the area is worthy of protection, and for the orders and prohibitions contained in the ordinance (Mannheim Administrative Court, Reference 5 S 251/91, UPR 93, p. 151). Under Article 33(3) BNatSchG the protection purpose of protected sites of habitats of wild fauna and flora and of bird species should also be designated in line with the respective conservation objectives for such sites (see section 3.1 above). Thus, the protection purpose shall include all conservation objectives relevant to the site; these will be the assessment criteria. When describing the protection purpose, it appears to be advisable to refer to the management plan describing in more detail the conservation objectives, thus avoiding the protection purpose and the ordinance as such to be overloaded with details (Gellermann 2004a, p. 29). Thus, the protection purpose described in the ordinance would be the framework for an eventual assessment, whereas the details could be regulated in a management plan.

Article 38(1) numbers (1), (2), (3), (4) and (5) define which orders and prohibitions could be included in the ordinance in accordance with Article 38(3) BNatSchG. Ordinances implementing provisions with regard to nature conservation could restrict, prohibit, or regulate any activity that negatively affects the protected area. Numbers (1) and (2) of Article 38(1) make clear that regulations of air traffic, shipping, internationally allowed military use, and scientific marine research may only be applied in accordance with international law.

The aim of number 3 Article 38(1) is to ensure that possible national restriction to fishing does not collide with the Common Fisheries Policy (CFP) and Council Regulation 2371/2002/EEC which was adopted to implement the CFP. Within the framework of the CFP, the EC has the competence to adopt provisions on the conservation, management, and exploitation of living aquatic resources. However, a collision with nature conservation legislation can be ruled out for legal reasons.[22] This is because according to Article 33 EC Treaty, the CFP primarily refers to the economic aspects of fisheries (Proelß 2004, p. 311) whereas the Member States are responsible for enforcing the protection of habitats at the European level under Article 174 EC Treaty in conjunction with Article 6 HD and Article 4 BD. Since the authorisation arising from Article 38 BNatSchG is restricted to implementing the protection of habitats at the European level only, the restrictions contained in the legal act to designate protected areas are always restrictions made for the purpose of nature conservation. Such measures are clearly not covered by the objectives enshrined in Article 33 EC Treaty, but by Article 174 EC Treaty (Gellermann 2004a, pp. 53ff; Schwarz 2004, pp. 18f). When competency is defined, the primary objective of the measure is important and not the possible side effects.[23] Measures for biodiversity protection in a specific SAC or SPA within a Member State's waters, which have certain effects on fishing, do not become a CFP measure just because of such effects (see Owen 2004, p. 64). The focus of the provision is of decisive importance. If such focus is on nature conservation, as is the case in this context, the measure is one of nature conservation (ECJ, Case C-164/97 para. 16, OJ 1999, p. 1139). Under Article 175(4) EC Treaty and – if the focus is on the protection of habitats at the European level – on the grounds of the explicit obligation arising from Article 6 HD and Article 4 BD, the Member States are responsible and have the competency for such measures. Member States can regulate human activities in NATURA 2000 sites that are negatively impacting marine species and habitats in order to protect or conserve these, without interference from the EU Fisheries Council even if the regulations have an impact or side effect on fisheries. The authors know very well that this is a highly disputed issue and will take this up in another publication.

Numbers 4 and 5 of Article 38(1) BNatSchG provide for restrictions on power generation from water, current and wind, and on the prospecting and extraction of mineral resources; that they are only admissible in accordance with Article 34 BNatSchG. On the grounds of the preceding, it can be concluded that the explorations and exploitations mentioned cannot be prohibited in advance for all explorations and exploitations, but only in those cases where it has been individually proven that they

[22] Such collision is also explicitly ruled out according to Article 13(1)(d) in conjunction with Article 14(2)(d) of the Draft Treaty Establishing a Constitution for Europe (6 August 2004 version) adopted by the European Council.

[23] ECJ, Case 70/88, para. 17, OJ I 1991, p. 4529; ECJ, Case C-164/97, para. 16, OJ 1999, p. 1139.

have adverse effects in accordance with Article 34 BNatSchG and that no exceptions are possible. The consequence is that the most relevant exploitations and those with a high potential for entailing damage can only be regulated to a very limited extent in the protective legal act (ordinance). However, it is permissible that the ordinance obligates the competent authorities to carry out an impact assessment for privileged exploitations and activities (Article 38(1) nr. 4 and 5) by way of an anticipated risk assessment (Gellermann 2004a, pp. 63ff). In connection with this, it can be necessary that the exploiters also carry out research and provide relevant information to the competent authorities.

The preference given to mining and wind power is rather astonishing. In terms of European legislation there is no reason to complain about the provision (Article 38(1) nr. 4 and 5) since it does not rule out completely the protection system foreseen under Article 6(2), (3) and (4) HD. This, however, is a new way followed by the legislator opposed to the protected areas on land. It is not the way to enhance biodiversity protection. Usually, appropriate orders and prohibitions directly regulate in advance possible activities and exploitations which might adversely affect protected areas. Additionally, the restrictions mentioned under numbers 1 to 5 of Article 38(1) should only serve to comply with the standards of international law and European legislation (justification of the law, BTD Case 14/6378 and 14/6878) and thus not the enhancement of economic issues. So far, the text of Article 38(1) numbers 4 and 5 BNatSchG is not covered by the justification of the law. In summary, it remains to be seen how far it will be possible for the legally required appropriate assessment in a particular case to permanently comply with the protection requirements under European legislation.

3.3 Nature conservation management

According to Article 6(1) HD the Member States shall establish, if need be, appropriate management plans specifically designed for the sites. According to Article 22(2) BNatSchG, protected areas include the respective management, development and restoration measures, or contain the necessary empowerments. At present, Germany is preparing for the protection of marine areas according to Article 38 BNatSchG by a combination of legal instruments (legal act – statutory ordinance) and technical and scientific instruments (management plans). The management plans shall contain detailed descriptions of the site, the environmental assets and the conservation objectives. Additionally, technical measures of conservation and restoration based on monitoring or scientific data shall also be mentioned.

References

Beyerlein U (2000) New Developments in the Protection of the Marine Environment: Potential Effects of the Rio Process. Zeitschrift für ausländisches öffentliches Recht und Völkerrecht, Bd. 55

Brandt E, Gaßner H (2002) Seeanlagenverordnung, Kommentar. Berliner Wissenschaftsverlag, Berlin

Bosecke T (2005) Vorsorgender Küstenschutz und Integriertes Küstenzonenmanagement (IKZM) an der deutschen Ostsee. Springer-Verlag, Berlin Heidelberg

Czybulka D (1996) Rechtspflichten des Bundes und der Länder zur Ausweisung und Beibehaltung von Schutzgebieten und Biotopen. In: Czybulka D (ed) Naturschutzrecht und Landschaftsplanung in europäischer Perspektive. Nomos Verlag, Baden-Baden, pp 39–53

Czybulka D (1999) Naturschutzrecht im Küstenmeer und in der Ausschließlichen Wirtschaftszone. Natur und Recht 21 (10):562–570

Czybulka D (2001) Geltung der FFH-Richtlinie in der Ausschließlichen Wirtschaftszone – Ein Urteil aus London und seine Folgen für das deutsche Naturschutzrecht. Natur und Recht 23 (1):19–27

Czybulka D (2001a) Das Rechtsregime der Ausschließlichen Wirtschaftszone (AWZ) im Spannungsfeld von Nutzungs- und Schutzinteressen, Zur Geltung des nationalen Rechts in der AWZ. Natur und Recht 23 (7):367–374

Czybulka D (2003) Der europäische und nationale Rechtsrahmen bei der Biotopvernetzung. In: Oldiges M (ed) Perspektiven des Naturschutzes. Nomos Verlag, Baden-Baden, pp 59-98

Czybulka D, Kersandt P (2000) Legal Regulations, Legal Instruments and Competent Authorities with Relevance for Marine Protected Areas (MPAs) in the Exclusive Economic Zone (EEZ) and the High Seas of the OSPAR Maritime Area. BfN-Skript, Band 27

Czybulka D, Kersandt P (2003) Rechtliche Rahmenbedingungen. In: Lozan J, Rachor E, Reise K, Sündermann J, Westernhagen Hv (eds) Warnsignale aus Nordsee und Wattenmeer, Wissenschaftliche Auswertungen Hamburg, pp 374 –383

Europäische Kommission (2000) NATURA 2000 – Gebietsmanagement: Die Vorgaben des Artikels 6 der Habitat-Richtlinie 92/42/EWG; Amt für amtliche Veröffentlichungen der Europäischen Gemeinschaften, Luxemburg

Gassner E, Bendomir-Kahlo G, Schmidt-Räntsch A, Schmidt-Räntsch J (2002) Bundesnaturschutzgesetz (BNatSchG), Kommentar, 2. Auflage, München

Gellermann M (1996) Rechtsfragen des europäischen Habitatsschutzes. Natur und Recht 18 (11/12):548–558

Gellermann M (2001) Natura 2000, Blackwell Wissenschafts-Verlag, 2. Auflage, Berlin

Gellermann M (2004) Recht der natürlichen Lebensgrundlagen in der Ausschließlichen Wirtschaftszone (AWZ) – dargestellt am Beispiel der Windkraftnutzung. Natur und

Recht 26 (2):75-81

Gellermann M (2004a) Nutzungsbeschränkungen in geschützten Meeresflächen im Bereich der Ausschließlichen Wirtschaftszone (AWZ) und des Festlandsockels, unveröffentlichter Endbericht zu einem Forschungsvorhaben des BfN

Graf Vitzthum W (1986) Die Europäische Gemeinschaft und das Internationale Seerecht; Archiv des Öffentlichen Rechts 111 (1):33–62

Ipsen HP (1979) EWG über See. In: Ipsen HP (ed) Recht über See. Festschrift für Rolf Stödter. Verlag Mohr Siebeck Hamburg, pp 167–207

Iven K (1998) Zur Praxis der Mitgliedstaaten bei der Ausweisung von Vogelschutzgebieten. Anmerkung zum Urteil des EuGH vom 19.05.1998 - Rs. C-3/96, Natur und Recht 20 (10):528–531

Janssen G (2002) Die rechtlichen Möglichkeiten der Errichtung von Meeresschutzgebieten in der Ostsee. Unter besonderer Berücksichtigung des deutschen und schwedischen Naturschutzrechts. Rostock Univ., Diss. 2001, Rostocker Schriften zum Seerecht und Umweltrecht, Bd. 19, Baden-Baden

Jarass HD (2002) Naturschutz in der Ausschließlichen Wirtschaftszone, Rostocker Schriften zum Seerecht und Umweltrecht, Bd. 17, Baden-Baden

Kersandt P (2002) Die Rechtslage für die Genehmigung von Offshore-Windenergieanlagen. In: Kumpfer W, Schlacke S (eds) Integrativer Umweltschutz, Schriftenreihe des Vereins für Umweltrecht, Bd. 13, Berlin, pp 119–130

Klinski S (2001) Rechtliche Probleme der Zulassung von Windkraftanlagen in der AWZ. Texte 62/01, Umweltbundesamt, Berlin

Lagoni R (2002) Die Errichtung von Schutzgebieten in der ausschließlichen Wirtschaftszone aus völkerrechtlicher Sicht. Natur und Recht 24 (3):121–133

Long R, Grehan A (2002) Marine Habitat Protection in Sea Areas under the Jurisdiction of a Coastal Member State of the European Union: The Case of Deep-Water Coral Conservation in Ireland. The International Journal of Marine and Coastal Law 17(2):235–261

Louis HW (2000) Bundesnaturschutzgesetz, Schapen Edition, 2. Auflage, Braunschweig

Nebelsieck R (2002) Die Genehmigung von Off-shore-Windenergieanlagen in der AWZ, Rechtsgutachten im Auftrag von NABU und WWF

Nordberg L (2003) Die Anwendbarkeit der Fauna-Flora-Habitat-Richtlinie und der Vogelschutzrichtlinie der Europäischen Gemeinschaft auf die marine Umwelt. In: Czybulka D (ed) Naturschutz und Rechtsregime im Küsten- und Offshore-Bereich – Tagungsband zum 4. Warnemünder Naturschutzrechtstag Nomos Verlag, Baden-Baden, pp 112–119

Owen D (2004) Protecting marine SACs and SPAs from fishing activities. Who has the power to impose restrictions? In: Ritterhoff J, Gubbay S, Zucco C (eds) Marine Protected Areas and Fisheries. BfN-Skript, Bd. 122, pp 61–65

Proelß A (2004) Meeresschutz im Völker- und Europarecht. Das Beispiel des Nordostatlantiks, Berlin

Rödiger-Vorwerk T (1998) Die Fauna-Flora-Habitat- Richtlinie der Europäischen Union und ihre Umsetzung in nationales Recht. Erich Schmidt Verlag, Umweltrecht Band 6, Univ., Diss, Berlin

Schumacher J, Fischer-Hüftle P (2003) Bundesnaturschutzgesetz, Kommentar.

Kohlhammer, Stuttgart
Schwarz KA (2004) Das maßgebliche Rechtsregime für Nutzungsbeschränkungen der Fischerei in der ausschließlichen Wirtschaftszone (AWZ)der Bundesrepublik Deutschland (Modul 1b). In: Gellermann M (ed) Nutzungsbeschränkungen in geschützten Meeresflächen im Bereich der Ausschließlichen Wirtschaftszone (AWZ) und des Festlandsockels (Annex), unveröffentlichter Endbericht zu einem Forschungsvorhaben des BfN
Spannowsky W (2000) Vorgaben der räumlichen Gesamtplanung für die Ausweisung besonderer Schutzgebiete; Umwelt- und Planungsrecht 20 (2):41–51
Ssymank A, Hauke U, Rückriem C, Schröder E (1998) Das europäische Schutzgebietssystem NATURA 2000. Bundesamt für Naturschutz, Bonn-Bad Godesberg
Stoll PT (1999) Meeresschutz im Küsten- und Offshore-Bereich im Hinblick auf nicht-stoffliche Einflüsse. Natur und Recht 21 (12):25–48
Stoll PT (2004) Völker-, europa- und nationalrechtliche Möglichkeiten der Beschränkung der Schifffahrt und wissenschaftlichen Meeresforschung im Bereich der ausschließlichen Wirtschaftszone und des Festlandsockels (Modul 1a). In: Gellermann M (ed) Nutzungsbeschränkungen in geschützten Meeresflächen im Bereich der Ausschließlichen Wirtschaftszone (AWZ) und des Festlandsockels (Annex), unveröffentlichter Endbericht zu einem Forschungsvorhaben des BfN
Wagner J (1990) Der Entwurf der Richtlinie „Fauna, Flora, Habitat" der Europäischen Gemeinschaft (88/C 247/03). Natur und Recht 10 (8):396–402
Weiß A (1999) Möglichkeiten der Regelung der Fischerei, des Bergbaus und der Schifffahrt in „Baltic Sea Protected Areas" (BSPAs) in der Ausschließlichen Wirtschaftszone (AWZ) der der Bundesrepublik Deutschland vorgelagerten Ostsee. BfN-Skript, Bd. 5
Wolf R (2004) Völker-, europa- und nationalrechtliche Möglichkeiten der Beschränkung der Verlegung von unterseeischen Kabeln und Rohrleitungen und Energieerzeugung aus Wasser, Strömung und Wind und Gewinnung und Aufsuchung von Bodenschätzen (Modul 1c). In: Gellermann M (ed) Nutzungsbeschränkungen in geschützten Meeresflächen im Bereich der Ausschließlichen Wirtschaftszone (AWZ) und des Festlandsockels (Annex), unveröffentlichter Endbericht zu einem Forschungsvorhaben des BfN, Verlag C Heymanns
Wolf R, Heugel M (2001) Stein-, Kies- und Sandabbau in der Ausschließlichen Wirtschaftszone bzw. auf dem Festlandssockel der Bundesrepublik Deutschland im Feld „OAM III", Rechtsgutachten im Auftrag des BfN
Wolff N (2004) Meeres- und Küstenbiodiversität. In: Wolff N, Köck W (eds) 10 Jahre Übereinkommen über die biologische Vielfalt. Nomos, Baden-Baden, pp 174–190
Wolfrum R (2001) Biodiversität – juristische, insbesondere völkerrechtliche Aspekte ihres Schutzes. In: Janich P, Gutmann M, Prieß K (eds) Biodiversität. Wissenschaftliche Grundlagen und gesellschaftliche Relevanz, Wissenschaftsethik und Technikfolgenb eurteilung, Band 10, Springer Verlag, pp 417ff
Wolfrum R (2002) Gutachterliche Stellungsnahme zur Problematik der Ausweisung von Meeresschutzgebieten in der AWZ im Zuge der Novellierung des BNatSchG (unpublished)

Part II: Site selection procedure

Chapter 3

Interpretation, identification and ecological assessment of the NATURA 2000 habitats "sandbank" and "reef"

Dieter Boedeker, Jochen C. Krause and Henning von Nordheim

German Federal Agency for Nature Conservation, branch office: Isle of Vilm

Abstract

In the German Exclusive Economic Zone (EEZ), the marine natural habitats *reefs* (NATURA 2000 code 1170) and *sandbanks which are slightly covered by sea water all the time* (1110) are found throughout the North Sea and the Baltic Sea. They serve several important functions, such as offering protection for rare and threatened species, as well as hosting important and threatened biotopes. Both habitat types occur in nearshore coastal waters as well as in offshore waters. Calculations show that with around 79% of all German North Sea sandbank habitats and with ca. 61% of those in the Baltic Sea, the majority of this habitat is situated in offshore waters of the German EEZ and the minority in the coastal waters of the Territorial Sea. The same situation applies for reefs in the North Sea (ca. 53% in the EEZ), whereas the majority of Baltic Sea reefs is situated closer to the coast (73% in the Territorial Sea). A great portion of the German EEZ sandbank habitats (64% North Sea, 99% Baltic Sea) and reef habitats (73% North Sea, 57% Baltic Sea) are included in the German offshore NATURA 2000 site nominations.

1 Introduction

According to the Habitats Directive of the European Union (EU), natural habitats of Annex I to the Directive, but also several animal and plant species of Annex II and IV are to be protected by EU Member States as Special Areas of Conservation (SACs). Together with the Special Protection Areas (SPAs) classified under the EU Birds Directive, they form the coherent European system of protected areas entitled NATURA 2000.

This article focuses on the identification and selection process of marine habitats in the German EEZ. The basic definitions and further mapping notes for all NATURA 2000 natural habitats are laid out in the *Interpretation Manual of European Union Habitats*. However, in the marine environment the application of the Habitats Directive has remained more difficult than originally expected, especially as regards the identification and selection of natural habitats in the offshore areas. EU-Member states had to face several gaps and deficiencies in the Interpretation Manual (e.g., imprecise habitat definitions) and in the applicability of the Habitats Directive for the offshore areas of the Member States as well (missing several threatened marine biotope types and species in the Annexes). To address these and other offshore marine issues, the EU Commission established in 2003 a marine expert working group (MEWG). According to the Terms of Reference of this group, the experts should propose, among other issues, new definitions of marine natural habitats, the best means of locating and assessing these habitats, and a site selection rationale.

Although preparatory work had been conducted for some years in Germany, the full national application of the Habitats- and Birds Directives in the German EEZ started only in 2002, directly after the legal basis was established through an amendment of the Federal German Nature Conservation Act.

In the following, methods that were applied for identifying and assessing offshore natural Annex I (Habitats Directive) habitats in the German EEZ are explained. Several scientific projects were carried out or commissioned by the German Federal Agency for Nature Conservation (BfN) under the umbrella of the *HabitatMareNATURA2000* project of BfN.

2 Definition of marine habitats in the German EEZ

Other than animal and plant species, Annex I habitats are not clearly determined by their names. For that reason the Scientific Working Group set up by the Habitats Committee of the European Commission issued the *Interpretation Manual of European Union Habitats* to further define and describe the habitats. The first version was published in 1994, and an amended version of 2003 anticipated the accession of 10 new EU Member States in 2004 (EC 2003). In the German EEZ, only two marine natural habitats are known to be present (Balzer et al. 2002):

- Sandbanks which are slightly covered by sea water all the time (NATURA 2000 Code 1110); and,
- Reefs (NATURA 2000 Code 1170).

The authors have only recently received information on existing gas vents in the German North Sea and Baltic Sea. Whether they belong to the natural habitat "Submarine structures made by leaking gases" (NATURA 2000 code 1180) could not yet be ascertained (Kudrass 2004, pers. comm., Schlüter 2005, pers. comm.).

2.1 Interpretation Manual of European Union Habitats

The basic definitions and further mapping notes for all NATURA 2000 natural habitats are laid out in the *Interpretation Manual of European Union Habitats* (EC 2003). It provides the following basic definitions for the natural habitats sandbank and reef relevant for the German EEZ:

Sandbanks which are slightly covered by sea water all the time

"Sublittoral[1] sandbanks, permanently submerged. Water depth is seldom more than 20 m below Chart Datum. Non-vegetated sandbanks or sandbanks with vegetation belonging to the *Zosteretum marinae* and *Cymodoceion nodosae*."

Reefs

"Submarine, or exposed at low tide, rocky substrates and biogenic concretions, which arise from the sea floor in the sublittoral zone but may extend into the littoral zone[2] where there is an uninterrupted zonation of plant and animal communities. These reefs generally support a zonation of benthic communities of algae and animal species including concretions, encrustations and corallogenic concretions.

[1] Sublittoral: zone between the low-tide mark and the edge of the continental shelf.

[2] Littoral zone: zone between the tide marks (also intertidal zone).

In northern Baltic areas, the upper shallow water filamentous algal-zone with great annual succession is normally well developed on gently sloping shores. *Fucus vesiculosus* is submerged at depth of 0.5–6 m in the sublittoral zone. A red algae zone occurs below the *Fucus* zone at depth of about 5 m to 10 m."

2.2 Application of the Interpretation Manual in Germany

In April 2002, the German Federal Nature Conservation Act (Bundesnaturchutzgesetz, BNatSchG) was amended. Article 38 of the Act established the legal basis and statutory necessity for the implementation of NATURA 2000 also in the EEZ of the North Sea and the Baltic Sea, areas that fall within federal jurisdiction. Before that time, the selection of German marine sites could be directed only within the Territorial Sea (up to 12 nautical miles), where the coastal states (Länder) have jurisdiction for the NATURA 2000 site selection.

The new legal situation led to a reconsideration of the national classifications of sandbanks and reefs in Germany (Ssymank et al. 1998) in order to adapt them to the specific conditions of offshore areas (EEZ). In anticipation of this change, as early as 2001 BfN began discussions with national and international marine experts. Within this process two scientific events can be highlighted:

1 In close cooperation with the European Commission (Nature & Biodiversity Unit, DG ENVIRONMENT), BfN conducted an European workshop on the *Application of NATURA 2000 in the Marine Environment* from 27 June to 1 July 2001 at the International Academy for Nature Conservation (INA) on the Isle of Vilm (Germany). By sharing experiences and expertise, this workshop aimed at guiding the implementation of the EU Habitats and Birds Directives in the marine environment, including discussions on definitions for natural marine habitats (Boedeker and von Nordheim 2002). One widely accepted general result of the workshop was that Annex I did not sufficiently reflect threatened marine habitats for EU waters. In addition, it was a common view that there was no scientific argument to restrict the eligible depth of a sandbank, and finally it was stated clearly that the wording in the Interpretation Manual (EC 2003), "seldom deeper than 20 m" does not actually exclude selecting sites for sandbanks below 20 m. In the North

[3] The Summary Record (in German) can be downloaded from: http://www.bfn.de/marinehabitate/en/downloads/berichte/Statusseminar_AWZ-Forschung_2002.zip

Sea, in particular, sandbanks or parts of it with high ecological value are known to occur in waters deeper than 20 metres.

2 In September 2002, BfN organised a national status seminar bringing together scientists who were commissioned to work on the implementation of NATURA 2000 in the German EEZ.[3] The participants discussed, among other things, the habitat definitions, and compiled more concrete applicable national characterisations and mapping notes for marine natural habitats.

The results of these workshops and the ongoing discussion within the MEWG formed the scientific basis for a reconsideration and modification of the German interpretations of sandbank and reef habitats, summarised in following text boxes.

Characteristic structural attributes for the identification of the natural habitat "Sandbanks which are slightly covered by sea water all the time"

- Sandbanks are sandy ridges that clearly rise above their surroundings.
- They must be permanently submerged and be mainly surrounded by deeper water.
- Their substrate is primarily a sand to gravel mix with a minimum thickness of 30 to 40 cm to provide habitats for typical sandy bottom communities, but patches of larger grain sizes, including boulders and cobbles may also be present on a sandbank as well as lower portions of mud.
- They are often free of vegetation, or only sparsely covered by macrophyte vegetation.
- Sandbanks can be found in association with mudflats and sandflats (1140) as well as with reefs (1170).
- Banks where sandy sediments occur in a layer over hard substrata are classed as sandbanks if the associated biota are dependent on the sand rather than the underlying hard substrata.

Characteristic structural attributes for the identification of the natural habitat "Reef"

- Reefs are submarine, or exposed at low tide, they rise mildly to prominently above the seafloor, and are topographically distinct from the surrounding seafloor, e.g., forming or emerging from a submarine sill, bank, slope or ridge.

- Geogenic reefs are characterised by benthic species that settle on hard compact substrata such as rocks or stones, moraine ridges with block, and stones surrounded by glacial till.
- Existing biogenic reefs in German waters include *Sabellaria* reefs[4] and mussel banks, that settle on solid and soft seabeds. Natural oyster beds were known to occur through to the beginning of the 20th century, but were extirpated (Nehring 2003).
- Reefs can be found in association with "sandbanks which are slightly covered by sea water all the time" (1110) and with mudflats and sandflats (1140).
- Reefs which are partly covered by mobile sediments should be classed as reefs if the associated biota are dependent on the rock rather than the overlying sediment.

3 Identification, ecological assessment and delimitation of natural marine habitats in the German EEZ

BfN used existing scientific expertise, including the results of previously concluded research projects (Rachor and Nehmer 2003, Gosselck et al. 1998), as well as thematic maps (e.g., Figge 1981, Nielsen 1992, LANU 1998, Geologische Karte MV 1994) for a preliminary assessment of the distribution of bottom sediments and of the marine species and communities in the German North Sea and Baltic Sea. Additionally, national and international red lists provided information on what degree species and biotopes were rare, declining or threatened (Riecken et al. 1994, von Nordheim et al. 1996, von Nordheim and Merck 1995, Merck and von Nordheim 1996, von Nordheim and Boedeker 1998).

In 2001, first sites of special ecological value in the North Sea and Baltic Sea and an initial distribution of marine Annex I natural habitats in the German Baltic Sea (Balzer et al. 2002) were published by BfN (Boedeker et al. 2001a, 2001b). This was the basis for new research within the *HabitatMareNATURA2000* project which started thereafter to determine the location of additional reefs and sandbanks, particularly in those areas

[4] Sublittoral habitat of the North Sea, in the Wadden Sea area threatened by complete destruction (Nordheim et al. 1996), not known in German offshore areas (Vorberg 1998).

where previous information gave first indications that such natural habitats were likely to be present. Therefore, in areas of special ecological value with insufficient information on sediments, additional analysis of the seafloor was carried out by the Institute of Geo-Science of Kiel University. The surveys included the following methods:

- hydro-acoustic mapping tools (multibeam, or sidescan sonar) to provide bathymetry and/or seabed character (see chapter 6); and,
- groundtruthing with a video sledge and/or seabed substratum sampling with grabs and corers.

Figure 1. All field work was supported by the SCUBA divers of the Federal Agency for Nature Conservation (photo BfN © Krause & Hübner)

In order to assess whether a morphologically recognisable bank-like structure would qualify as sandbank or reef habitat, it was further necessary to examine its biology. Existing biological data, particularly for the benthos, was amassed to identify the presence of red listed species and biotopes. In case no historical or actual information on the distribution of benthic species or communities existed, specific projects commissioned to the Baltic Sea Research Institute, Warnemünde and the Alfred Wegener – Institute, Bremerhaven provided further expert assessment (see chapters 7 and 8).

Finally, all relevant spatial data were summarised and transferred into the Geographic Information System (GIS) at BfN. Biological data were overlaid upon physical data with the goal of identifying sandbank and reef habitats.

The following sections, 3.1 and 3.2, contain detailed descriptions of the scientific steps that were undertaken in order to map these marine Annex I natural habitats in the German EEZ.

3.1 Identification, ecological assessment and delimitation of sandbank habitats

The selection criteria for Annex I habitats are given in Annex III of the Habitats Directive (see chapter 4). Before these criteria could be applied, BfN had to identify the almost complete suite of potential sandbanks in the German EEZ. Thus, the following steps were undertaken:

1 A 3D-Model of the seafloor was produced by using a TIN (Triangular Irregular Network) and a GIS (see chapter 5). In such a way, structures which rise above the sea floor, like sandbanks, could be recognised more easily than just from two-dimensional charts. The model delivered also necessary information on morphological properties of submarine structures, such as levelness, bank character, exposure gradient and orientation which could be queried from GIS. From these, several banks could be identified within the German offshore area as basis for the next steps. Because the scale of the different data sets varied widely (between 1:20,000 and 1:500,000), the spatial resolution of the different modelled layers also varied. Therefore, a minimum size of 1 km^2 had to be set as a threshold for any area to be identified as a bank (see chapter 5).

2 BfN overlaid the resulting modelled layer of banks with various sediment data sets from grab samples, sidescan sonar data, and digital geological maps. With this overlay a draft set of sandy banks was produced which however did not yet contain any information on their biology.

3 Finally, (sandy) bank polygons were matched with available information on benthic infaunal species (see chapters 7 and 8) and, as were available, marine benthic biotope types according to Riecken et al. (1994). Additional BfN data came from recent underwater video evaluations of all potential sandbanks, grab sampling and/or observations by SCUBA dives.

Figure 2 illustrates one such GIS map overlay for the easternmost part of the German Baltic Sea. The model layers shown in the figure are:

- boundaries of potential sandbanks (see chapter 5), i.e., banks according to the result of the TIN model compilation (green lines);
- location of grab samples where the marine amphipod *Bathyporeia* sp. was found (see chapter 8). Its habitat in the Baltic Sea is restricted to fine to medium sand in the shallow sublittoral (black points);
- qualifying sandbanks, representing the overlap of *Bathyporeia* sp. and the modelled bathymetric banks (yellow shaded polygons).

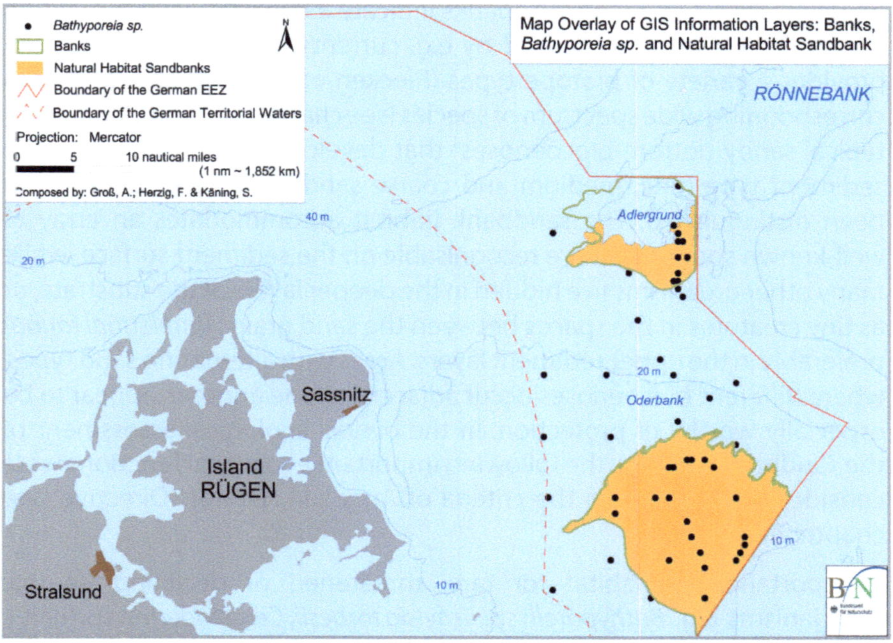

Figure 2. Map overlay of GIS information layers: structural bank boundaries, occurrence of the amphipod *Bathyporeia* sp. and polygons of the natural habitat sandbank for the *Adler Ground* and the *Odra Bank* in the German EEZ of the *Pommeranian Bay*

The two areas which thus qualify as sandbanks are the *Adler Ground* (*Adlergrund*) and the *Odra Bank* (*Oderbank*). The *Odra Bank* is a pure sandbank and the structural bank boundary is the same as what was identified with the overlay; whereas for the *Adler Ground*, its upper parts qualify as natural reef habitat, but not as sandbank. According to Zettler (pers. comm. 2005), the abundances of *Bathyporeia* were higher on the sandbank habitats of *Odra Bank* and *Adler Ground* compared to the shallower and more level sea bed in between both sites.

In summary, it was found that sandbank habitat type as at Annex I of the Directive is spread throughout the North Sea and Baltic Sea. It occurs in inner marine waters, in nearshore coastal waters, as well as further offshore in the EEZ. In the North Sea, it is frequently associated with sand flats, with the outer parts of large estuaries, as well as with reefs. Ripple

[5] Biocoenoses: associations of organisms living together in a certain habitat.

markings found on many sandbanks indicate a dynamic environment on the top of these banks caused by e.g. currents. The substrate diversity provides a variety of biotope types (Riecken et al. 1994) and, in turn, a corresponding wide spectrum of species (see chapters 7 and 8). Numerous typical sandy bottom biocoenoses[5] that develop interdependently with sediment type (fine, medium and coarse sand) and water depth have been distinguished. The sandbank habitat accommodates an array of well-known species that are recognisable on the sediment surface, while many other organisms live hidden in the deeper layers of the substrate, or as tiny creatures in the spaces between the sand grains (*interstitial fauna*) preferably in the upper sediment layers. Areas with alternating sand types, where different biocoenoses occur adjacent to one another, appear to be especially worthy of protection. In the overall ecological assessment of the sandbank habitats, the following important ecological functions were considered according to the criteria of Annex III Habitats Directive (see chapter 4):

- importance as habitat for rare, threatened or declining benthic organisms, e.g., *Bathyporeia* sp., *Travisia forbesii*, *Cerastoderma* sp. (Baltic Sea), *Spisula* sp., *Glyceria* sp., and *Lagis koreni* (North Sea);
- importance as stepping-stone for the expansion of bottom organisms into other parts of the seas and thereby preventing isolation and genetic depletion of populations (e.g,. the *Doggerbank*);
- importance as regeneration and recolonisation reservoir after catastrophic oxygen depletion events with mass-mortality in deeper marine areas or similar catastrophes during iced winters in nearshore shallow waters, (e.g., *Odra Bank* and *Adler Ground*);
- importance as feeding, resting and nursery grounds for demersal fish species and marine mammals; and,
- importance as feeding habitat for resting and wintering seabirds, e.g., sea ducks (incl. common scoter) and loons.

The above assessment resulted in a set of sandbank habitats found to be worthy of protection, and led to the following areas being selected for protection according to their representativity, conservation status and value, restoration possibilities and size:

For the North Sea:

- *Borkum Reef Ground* (*Borkum-Riffgrund*) with bank like elevations of the shelf and high sediment dynamics;
- *Amrum Bank* with representative parameter values for this natural habitat (figure 3);
- *Doggerbank* with glacial relicts.

Figure 3. *Amrum Bank* (North Sea) with sea stars (*Asterias rubens*) on its surface (photo BfN © Krause & Hübner)

Figure 4. *Fehmarn Belt* (Baltic Sea), top of a subaquatic dune (sandbank subtype megaripple) with sea star (photo BfN © Krause & Hübner)

For the Baltic Sea:

- *Fehmarn Belt* with dynamic subaquatic dunes (figure 4);
- *Adler Ground* with bank-like elevations; and
- *Odra Bank* as a submerged former dune.

Boundaries were generally drawn at the transition from the slopes of the bank into the surrounding plains and/or at the transition of typical sandbank biotopes to other non-typical biocoenoses. Boundaries which ended up in shallow or coastal areas were demarcated by a straight line.

3.2 Identification, ecological assessment and delimitation of reef habitats

The guiding principles for the identification and pre-selection of reef habitats were in principle similar to those for sandbanks:

1 Potential reefs were identified by interpretation of thematic maps and the analysis of scientific data and literature reviews. In order to achieve a comprehensive suite of potential reef sites in the German EEZ, all locations with glacial tills, biogenic hard substrates (e.g., mussel beds), and fields of boulders and blocks along submarine moraine ridges were incorporated into the GIS. Figure 5 illustrates such a GIS-supported analysis. On the basis of a sediment distribution map, those areas which indicated hard bottom substrates were chosen for additional closer investigations by sidescan sonar surveys (green profiles). The red shaded polygons show the resulting "potential reefs" (not all sidescan sonar profiles and no video profiles are shown in the figure).

2 In those identified potential reef areas where no field work had yet been done, the following research approach was taken:

- assessment of grain sizes of cobble fields with underwater video recordings (laser measurement), and in some areas, scientific diving;
- ascertainment of biological features through bottom dredge trawls (benthic samplings), and/or video profiles, and in some areas scientific diving;
- photo and video documentation of habitat types by video profiling and scientific diving.

3 The last step was to achieve a complete suite of ecologically valuable reefs which also reflect, in a representative way, the different ecological forms and features of the habitat types according to Annex III of the Habitats Directive. The following forms of reef habitats occur in the German EEZ:

North Sea:

- Reefs in the form of boulder or cobble fields, which arise from the seafloor (typical for the central part of *Sylt Outer Reef* (*Sylter Außenriff*), figure 6).
- Stony reef bands along the slope of the glacial Elbe Valley (*Sylt Outer Reef*, figure 7).
- Scattered stony reefs (*Borkum Reef Ground*).

Baltic Sea:

- Stony reefs and mussel beds at the slopes of e.g., *Fehmarn Belt* with high salinity (up to 25 psu) and with macrophyte vegetation (figure 8); *Kadet Trench* (*Kadetrinne*) representing "deeper reefs" with a medium salinity of 10 to 18 psu (figure 9).

- Reefs in the form of boulder or cobble fields, which arise on thetop of a shallow bank e.g., (*Adler Ground*, with low salinity and rich macrophyte vegetation (figure 10).
- Stony reefs and mussel beds on a deeper bank with low salinity and without macrophyte vegetation (*Rønne Bank* (*Rönnebank*)).

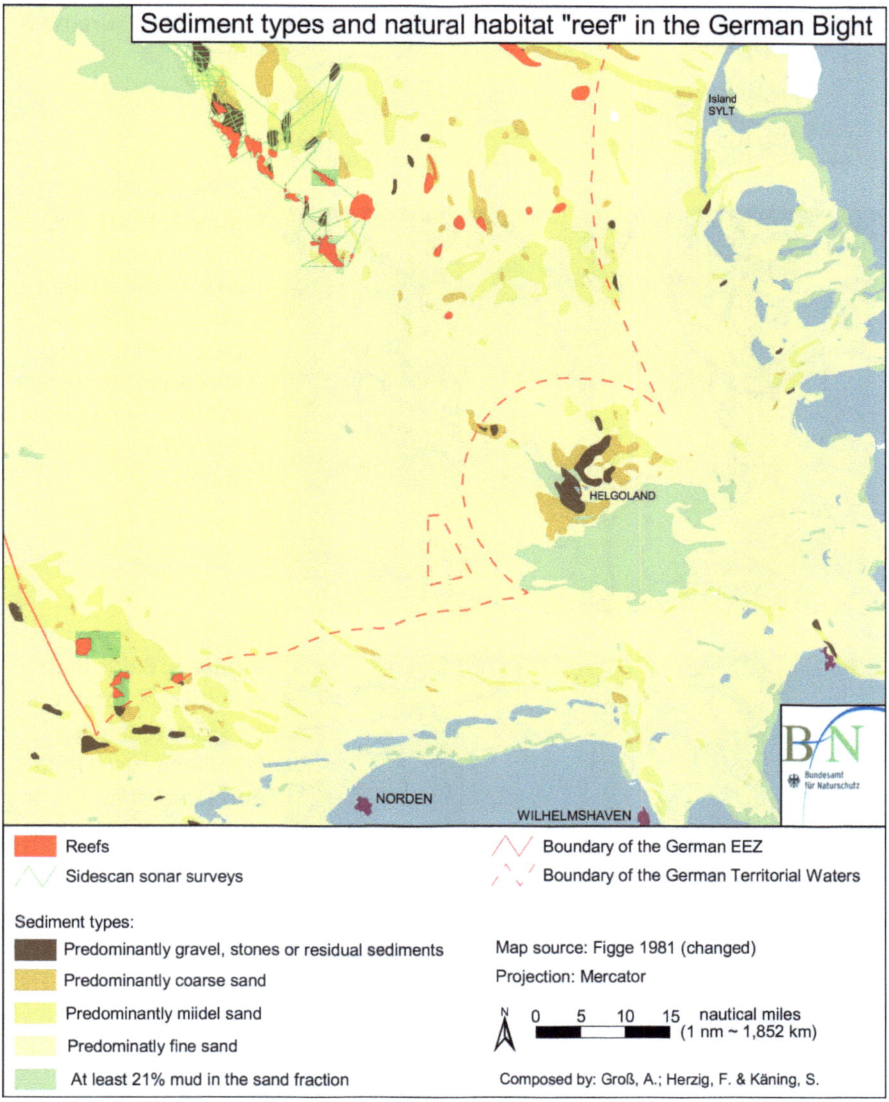

Figure 5. Sidescan sonar survey as basis for the identification of the natural habitat reef. Sediment data modified from data provided by BSH

Figure 6. Typical species on top of a North Sea stony reef (e.g., *Alcyonium* sp., *Pomatocerus triqueter*, *Flustra* sp. and *Cionia* sp.)at *Sylt Outer Reef* (photo BfN © Krause & Hübner)

Figure 7. North Sea stony reef at the slopes of the ancient glacial Elbe Valley with typical epifauna (e.g., *Pomatocerus triqueter* and *Metridium senile*) (photo BfN © Krause & Hübner)

Figure 8. Baltic high-salinity reef community on stony reef (e.g., *Laminaria saccharina* and *Dendrodoa* sp.) at the slope of *Fehmarn Belt* (photo BfN © Krause & Hübner)

Figure 9. Baltic medium-salinity reef community on stony reef (with e.g., Hydrozoa, *Laminaria saccharina* and *Ctenolabrus rupestris*) at the slopes of *Kadet Trench* (photo BfN © Krause & Hübner)

Figure 10. Baltic low-salinity reef community (*Fucus serratus* – stands with *Gobius flavescens*) on top of *Adler Ground* (photo BfN © Krause & Hübner)

4 Spatial distribution and protection of natural marine habitats in the German North Sea and Baltic Sea

In the German EEZ, sandbanks and reef habitats have been thoroughly investigated within the BfN research programme *HabitatMare-NATURA2000* since 2002, as a substantial scientific deepening of the preliminary works in the 1990s. It became clear that these habitats are spread throughout the German North Sea and Baltic Sea. They occur in inner marine waters, in nearshore coastal waters, as well as further offshore, in the EEZ. Table 1 provides data on the spatial extend of the two natural habitat types in the German North Sea and Baltic Sea. Figure 11 illustrates the distribution of the natural habitats sandbank and reef in the German EEZ of the North Sea and Baltic Sea.

Table 1. Sandbank and reef habitats in the German North Sea and Baltic Sea (numbers indicate area covered in hectares)

Natural habitat	**North Sea** (EEZ+ territorial Sea)	**Baltic Sea** (EEZ+ territorial Sea)	**North Sea** (EEZ)	**Baltic Sea** (EEZ)
Sandbanks total	438,555	94,195	347,995	57,129
Sandbanks inside pSCIs[6]	273,190	69,143	223,190	57,094
Reefs total	45,283	171,996	24,081	46,377
Reefs inside pSCIs	24,560	40,772	17,627	26,459

[6] pSCI: proposed Site of Community Importance

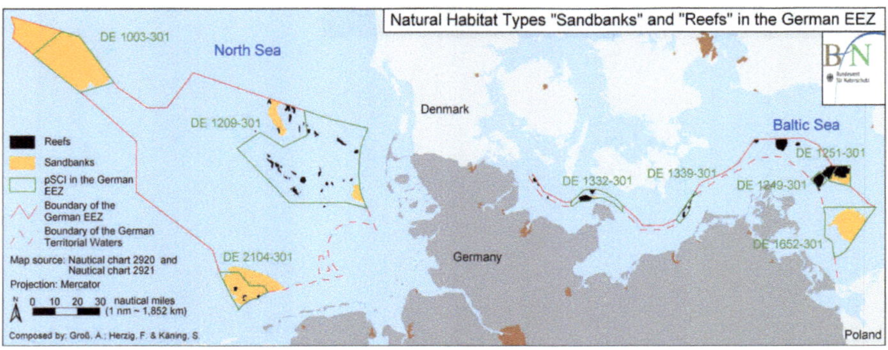

Figure 11. Occurrence of the Natural Annex I Habitats "Sandbanks which are slightly covered by sea water all the time" (NATURA 2000-Code 1110) and "Reefs" (NATURA 2000-Code 1170) in the EEZ of the German North Sea and Baltic Sea

The source for the number values in table 1 is the BfN NATURA 2000-database, which stores data from the coastal "Länder" (States) and data collected by BfN itself. The calculations show that with around 79% of all German North Sea sandbank habitats and with ca. 61% of those in the Baltic Sea, the majority portion of this habitat is situated in the German EEZ, whereas the minority is found in the coastal waters of the German Territorial Sea. The same situation applies for reefs in the North Sea (ca. 53% in the EEZ), whereas the majority of Baltic Sea reefs (73%) is situated closer to the coast in the Territorial Sea.

The research results indicate that sandbank and reef habitats serve several important ecological functions, such as offering protection for rare and threatened species as well as providing breeding, nursery, feeding and resting habitats. Furthermore, they host important biotope types and communities of marine animal and plant species. Areas with alternating substrates where different biocoenoses occur adjacent to one another appeared to be especially worthy of protection. Consequently, a great portion of the sandbank habitats (64% North Sea, 99% Baltic Sea) and reef habitats (73% North Sea, 57% Baltic Sea) are comprised in the pSCIs and thus are nominated as NATURA 2000 Marine Protected Areas (MPAs) (see chapter 4).

References

Balzer S, Boedeker D, Hauke U (2002) Interpretation, Abgrenzung und Erfassung der marinen und Küsten-Lebensraumtypen nach Anhang I der FFH-Richtlinie in Deutschland. Natur und Landschaft, 77 (1):20–28

Boedeker D, von Nordheim H (2002) Application of NATURA 2000 in the Marine Environment. BfN-Skripten 56. Bundesamt für Naturschutz, Bonn. http://www.bfn.de/09/natura2000marin.pdf

Boedeker D, Herrmann U, Hübner P, Kias M, Merck T, Rachor E, von Nordheim H (2001a) Ökologisch besonders wertvolle marine Gebiete im deutschen Nordseebereich. BfN-Map: 2. überarbeitete Auflage http://www.bfn.de/marinehabitate/en/downloads/Nordsee.pdf

Boedeker D, Herrmann U, Hübner P, Kias M, Meißner J, von Nordheim H (2001b) Ökologisch besonders wertvolle marine Gebiete im deutschen Ostseebereich. BfN-Map: 2. überarbeitete Auflage http://www.bfn.de/marinehabitate/en/downloads/Ostsee.pdf

EC (2003) Interpretation manual of European Union habitats. EUR 25. European Commission (DG Environment), Brussels

Figge K (1981) Sedimentverteilung in der Deutschen Bucht – Karte Nr. 2900, Deutsches Hydrographisches Institut, Hamburg

Geologische Karte MV (1994) Geologische Karte Mecklenburg-Vorpommern 1:500.000. Geologisches Landesamt Mecklenburg-Vorpommern

Gosselck F, Bönsch R and Kreuzberg M (1998) Fachgutachten über wissenschaftliche Grundlagen zur Ausweisung und zum Management mariner OFFSHORE-Schutzgebiete im Bereich der Hoheitsgewässer und der AWZ Deutschlands in der Ostsee und deren Integration in das System von "Baltic Sea Protected Areas". FuE-Vorhaben im Auftrag des Bundesamtes für Naturschutz

LANU (Landesamt für Natur und Umwelt) (Hrsg.) (1998) Geologische Karte von Schleswig-Holstein 1:500.000. Flintbek

Merck T, von Nordheim H (1996) (eds) Rote Listen und Artenlisten der Tiere und Pflanzen des deutschen Meeres- und Küstenbereichs der Ostsee. Schriften-reihe für Landschaftspflege und Naturschutz, Heft 48

Nehring S (2003) Gebietsfremde Arten in den deutschen Gewässern – ein Risiko für die Biodiversität. Alien species in German waters – a risk for biodiversity. Schriftenreihe des BMVEL „Angewandte Wissenschaft", Heft 498:40–52

Nielsen PE (ed) (1992) Map on Bottom Sediments around Denmark and Western Sweden. Geological Survey of Denmark Kortserie nr. 38; Geological Survey of Sweden. Serie Ba nr. 48

Rachor E, Nehmer P (2003) Erfassung und Bewertung ökologisch wertvoller Lebensräume in der Nordsee. Abschlussbericht für das F+E-Vorhaben FKZ 899 85 310 (Bundesamt für Naturschutz). Alfred-Wegener-Institut für Polar- und Meeresforschung, Bremerhaven

Riecken U, Ries U, Ssymank A (1994) Rote Liste der gefährdeten Biotoptypen der Bundesrepublik Deutschland. Schriften-reihe für Landschaftspflege und Naturschutz, Heft 41

Ssymank A, Hauke U, Rückriem C, Schröder E (1998) Das europäische Schutzgebietssystem Natura 2000. BfN-Handbuch zur Umsetzung der Fauna-Flora-Habitat-Richtlinie (92/43/EWG) und der Vogelschutzrichtlinie (79/409/EWG). – Schriftenreihe für Landschaftspflege und Naturschutz 53

von Nordheim H, Boedeker D (eds) (1998): Red List of Marine and Coastal Biotopes and Biotope Complexes of the Baltic Sea, Belt Sea and Kattegat – Including a comprehensive description and classification system for all Baltic Marine and Coastal Biotopes. HELCOM-Baltic Sea Environ-ment Proceedings No. 75. Helsinki Commission

von Nordheim H, Norden Andersen O, Thiessen J (eds.) (1996) Red Lists of Biotopes, Flora and Fauna of the Trilateral Wadden Sea Area 1995. Helgol. Meeres-untersuchungen. 50 (suppl.)

von Nordheim H, Merck T (eds) (1995) Rote Liste der Biotoptypen, Tier- und Pflanzenarten des deutschen Wattenmeer- und Nordseebereichs. Schriften-reihe für Landschaftspflege und Naturschutz, Heft 44

Vorberg R (1998) Sandkorallenriffe. In: Umweltatlas Wattenmeer Band I. Nordfriesisches und Dithmarscher Wattenmeer. Landesamt für den Nationalpark Schleswig-Holsteinisches Wattenmeer, Umweltbundesamt (Hrsg.) – Ulmer. 96–97

Chapter 4

Rationale behind site selection for the NATURA 2000 network in the German EEZ

Jochen C. Krause, Dieter Boedeker, Ina Backhausen, Kathrin Heinicke, Annette Groß and Henning von Nordheim

German Federal Agency for Nature Conservation, branch office: Isle of Vilm

Abstract

In the offshore waters of the European Seas, the selection of NATURA 2000 sites has only recently started. The guidelines for this process are still being developed by a Marine Expert Working Group under the umbrella of the European Commission. In 2002, following the adoption of the new German Federal Nature Conservation Act (Bundesnaturschutzgesetz), a relevant survey programme of the German Federal Ministry for Environment, Nature Conservation and Nuclear Safety and the German Federal Agency for Nature Conservation with the title *HabitatMareNATURA2000* started. Its purpose was to identify, locate and assess Annex I habitats and Annex II species of the Habitats Directive (92/43/EEC), and Annex I-bird species as well as regularly occurring migratory birds of the Birds Directive (79/409/EEC) in the offshore area of the German North Sea and Baltic Sea. The present chapter outlines how scientific conclusions (summarised in the more technical chapters of parts III and IV of this book) were used for this task in the German Exclusive Economic Zone (EEZ). For habitats and species, the selection process itself was performed according to the criteria given in Annex III (so-called stage I criteria) of the Habitats Directive and according to Article 4 of the Birds Directive, respectively. As a result of this programme, Germany nominated to the European Commission in 2004 a comprehensive set of ten new marine NATURA 2000 sites in the German EEZ of the North Sea and the Baltic Sea: eight under the Habitats Directive and two under the Birds Directive. It was the first Member State to do so. Short descriptions of the habitats and species of each site are given in this chapter. In total, 31.5% (1,040,783 hectares) of the German EEZ is now covered by NATURA 2000 sites. This shows that despite many commonly recognised obstacles (i.e., limited information on sea-bottom habitats and communities, or missing necessary scientific research programmes to fill knowledge gaps), it was possible to select and nominate sites in offshore areas which fulfil the quality objectives of the European Nature Directives. However, the overall conservation value of the selected Marine Protected Areas (MPAs) within the EU still depends on the collective efforts of all Member States and, particularly, on the implementation of effective management plans.

1 Introduction

The establishment of MPAs is generally motivated by a desire to conserve and maintain biodiversity, or to protect and sustain stocks of harvested species for their future use (e.g., Ray 2004). Quite often, both objectives are mixed in regional MPA debates, with varying success (Kaiser 2004). In the European context, however, it is mainly the conservation of biodiversity that drives the creation of a coherent network of protected areas in the terrestrial and marine environment. This is stated in Article 2 of the Habitats Directive (92/43/EEC): *"Ensuring biodiversity through the conservation of natural habitats and of wild fauna and flora in the European territory of the Member States to which the Treaty applies".*

Both the Habitats Directive and the Birds Directive (79/409/EEC) have regulations on protected areas which form the basic requirements of the NATURA 2000 network on land and sea. These two EU Directives regulate the specific actions to be carried out by the Member States for the identification, selection, examination and nomination of protected areas, and they name fundamental principles for their management. For the development of the NATURA 2000 network, site selection and the implementation of management measures in the selected areas are divided into two consecutive steps. This is in contrast to a simultaneous approach often proposed by nature conservation literature (e.g., Agardy 1997). The experience with implementing NATURA 2000 in the last decade indicates that the separate development of management plans for MPAs is not a disadvantage per se. Past decisions of the European Court (see chapter 2) underlined that the selection of sites of ecological importance for European wild flora and fauna may not be influenced by economic concerns[1] and supported a break from the ignominious tradition of identifying protected areas mainly in areas of low economic interest. As a consequence, selection steps, which in the past were left more or less unnoticed in the hands of scientists and conservationists, need now to be well documented so that each selected square metre can be justified to relevant stakeholders. Such increased accountability as well as increased opposition against the selection of areas of economic values seem to be the main issues that prolonged the implementation process for the terrestrial areas that, actually, should have been finalised already in 1998 according to Article 4(3) of the Habitats Directive.

The selection of marine NATURA 2000 sites in the EEZ began in Germany after the revision of the German Federal Nature Conservation Act in 2002, in accordance with the opinion of the EU Commission (EC letter, 1998) and supported by an English High Court decision in 1999 (1336/1999). The latter stated that both the Habitats and the Birds Directive apply to the EEZs of the EU Member States. Since marine research requires sophisticated and often expensive research techniques, additional delays arose due to the enormous extension of the seafloor, which, in contrast to European terrestrial areas, remains poorly studied, mapped, and inventorised.

An essential guide for the information required for the selection of sites was given by the Commission decision 97/266/EC, which summarises the minimum information requirements for the NATURA 2000 data forms. The decision points out that all information of Annex I species of the Birds Directive as well as Annex I habitats and Annex II species of the Habitats Directive must be included in the EU data forms. Information on additional but unlisted fauna and flora is desirable but not necessary.

This chapter outlines how Germany applied the available scientific information achieved by the MPA projects that are presented in the following chapters of this book. Finally, a set of ten new MPAs in the German EEZ was nominated in May 2004 to the European Commission as a result of this selection process. To date, Germany is the only Member State of the European Union that has nominated such a comprehensive list of MPAs in its EEZ as part of the NATURA 2000 network.

---

1 However, socio-economic interests are considered in Article 6 of the Habitats Directive, which regulates management as well as assessment and permission for plans and projects of the selected sites.

2 Assessment of habitats and species for the identification and location of NATURA 2000 sites

2.1 Assessing Annex I habitats

Of those habitats listed in Annex I of the Habitats Directive (92/43/EEC), only two offshore habitats exist in the German EEZ:

- reefs (Code 1170), and
- sandbanks slightly covered by seawater all the time (Code 1110).

Methods engaged for the identification and location of Annex I habitats in the German EEZ are described in chapter 3 based on the results which are explained in more detail in chapters 5, 6, 7 and 8 of this book.

At the national level, the relative importance of identified sandbanks and reefs were assessed using the four *stage I criteria* of the Habitats Directive (Annex III):

a degree of *representativity* of the natural habitat type on the site;
b *area* of the site covered by the natural habitat type in relation to the total area covered by that natural habitat type within the national territory;
c degree of *conservation of the structure and functions* of the natural habitat type concerned and *restoration* possibilities; and
d *global assessment* of the value of the site for conservation of the natural habitat type concerned.

Representativity (criterion a) was assessed for the German marine habitat sites by further subdividing the German marine biographical regions using abiotic criteria such as depth and substrate type, and biotic criteria such as distinct benthic communities. This was to ensure that the selected habitat sites represent the various benthic species and communities in different regions of the North Sea and the Baltic Sea (see chapters 7 and 8).

For the stage I criterion *area* (criterion b), a relevant proportion of the area covered by the habitat type of the selected site was chosen. However, given the normally smooth transitions and most often missing sharp boundaries between various habitats and communities on the seafloor, accurate area measurements and statements on the extent of any given habitat type are difficult. These difficulties are recognised in the EU Decision 97/266/EC whereby data may be summarised into three broad classes (class A: $100\%>p\geq15\%$, class B: $15\%>p\geq2\%$ and class C: $2\%>p$, with p=measured percentage). In the German selection process, the total size of the habitat plus the estimated error of the assessment fit into one of the classes without problems. Thus, finally, 64% of the sandbanks and 73% of the reefs in the German North Sea EEZ, and 99% of the sandbanks and 57% of the reefs in the German Baltic Sea EEZ were selected. However, these nominations cover only 51% of the sandbanks and 39% of the reefs of the total German North Sea area (i.e., Territorial Waters and EEZ), and 61% of the sandbanks and 15% of the reefs in the total German Baltic Sea area, respectively (see chapter 3, section 4).

Are these percentages enough to reach the targets of the European Habitats Directive? In biogeographical seminars, the EU Commission, together with non-governmental organisations and independent experts invited by the Commission, evaluates the sites proposed by the Member States. As a non-strict numerical mechanism for deciding on the sufficient or insufficient level of representation of habitats in sites selected, the following predetermination percentages have been employed in such evaluations (Boillot et al. 1997). A proposal that covers less than 20% of the feature in need of selection according to the Directive would normally be considered inadequate. Where more than 60% is covered, it would normally be considered sufficient. For proposals that cover between 20% to 60% of a feature, the conclusion as to whether the Directive requirements are met or not would be based on independent expert judgement from seminars on the habitat or species concerned. Therefore, based on these guidelines by Boillot et al. (1997) – and expecting further nominations of Special Areas of Conservation (SACs) including sandbanks and reefs in the German Territorial Waters[2] – the German nominations are considered to adequately satisfy the requirements of the Directive.

As to the *structural and functional requirements* of a particular habitat (criterion c): clearly, the geological structure of sandbanks and reefs is not a conservation value in and of itself. Additional criteria were considered: ecological ones, such as the various zoobenthic[3] or phytobenthic[4] community types; and conservation criteria, such as the Red List categorisation and the number of threatened species present. Preference was given to sites with a high proportion of rare and threatened benthic species (Red List species) or communities (von Nordheim and Merck

---

2 In the German Territorial Sea (12-nautical mile zone), the German States (Länder) are responsible for the nomination of NATURA 2000 sites. The overall nomination of Germany in the North Sea and the Baltic Sea will be assessed by the EU Commission. The nomination process was not yet completed in Germany at the time of publication.

3 Zoobenthic: animals living in or connected to the seafloor.

4 Phytobenthic: plants living in or connected to the seafloor.

1995, Merck and von Nordheim 1996). In addition, sites were assigned a higher conservation value when species and habitats of community interest are ecologically interlinked, such as reefs and sandbanks (which are important foraging sites for migrating birds). In general, the possibility of active *habitat restoration* of sites in the German EEZ was considered extremely difficult, if not, implausible. Natural recovery times from identified disturbances (e.g., sand and gravel extraction) were estimated from expert opinions.

Finally, criteria a to c were summarised and integrated in the final *global assessment* (criterion d). The summary and integration was done with additional nature conservation information from the relevant research projects (i.e., this book) and from common scientific knowledge (e.g., the *Odra Bank* (*Oderbank*) in the Baltic Sea as a likely feeding ground for Baltic sturgeons after their successful reintroduction).

2.2 Assessing Annex II species

The species listed in Annex II of the Habitats Directive currently represent only a small share of the threatened species in the European Seas. For example, only anadromous fish species[5] are listed in Annex II. With no full list of marine fishes of concern and no marine invertebrates, the Habitats Directive appears to have been established biased towards terrestrial conditions.

Further, Article 4(1) of the Habitats Directive complicates the selection process by stipulating that for aquatic species that range over wide areas, a site for the NATURA 2000 network (SAC) can be proposed only if it provides "*physical and biological features essential to their life and reproduction*." Whether SACs could be identified at all for such wide-ranging species given these stringent criteria, was positively answered by an ad hoc meeting convened by the European Commission (EC 2001). With regard to harbour porpoises (*Phocoena phocoena*), it was concluded that those areas would be identifiable on the basis of:

- continuous or regular presence of the species (although subject to seasonal variations);
- good population density (in relation to neighbouring areas);
- high ratio of young-to-adults during certain periods of the year.

In the German EEZ, three marine mammal species and six anadromous fish species listed in Annex II occur regularly (table 1). Due to low salinity

[5] Anadromous fish species: a fish species that migrates, as an adult, from the sea to freshwater to spawn.

in the eastern parts of the Baltic Sea, some Annex II freshwater fish species may also use offshore seabeds as feeding grounds.

Table 1. Annex II species regularly occurring in the German EEZ of the North Sea and the Baltic Sea

EU Code	Common Name	Scientific Name
Mammals		
1351	Harbour porpoise	*Phocoena phocoena*
1364	Grey seal	*Halichoerus grypus*
1365	Harbour seal	*Phoca vitulina*
Fish		
1099	River lamprey	*Lampetra fluviatilis*
1095	Sea lamprey	*Petromyzon marinus*
1101	Baltic sturgeon, Atlantic sturgeon	*Acipenser oxyrinchus oxyrinchus, A. sturio*
1102	Allis shad	*Alosa alosa*
1103	Twaite shad	*Alosa fallax*

Due to the stringent criteria of Article 4(1), the available data for most Annex II species were not sufficient to identify and demarcate sites which qualify for the nomination as proposed Site of Community Importance (pSCIs)[6] in the German EEZ. However, for sites selected due to the presence of Annex I habitats, often Annex II species also occur; these Annex II species have to be protected in the same way as in the areas exclusively selected for them.

In the German North Sea, sites qualifying as pSCIs could be identified only for two Annex II species: harbour porpoise (see chapter 11) and twaite shad (*Alosa fallax*) (Stelzenmüller et al. 2004). Analysis of long-term historical catches of twaite shads reveals concentrations of adults in the coastal and estuarine areas of the rivers Elbe and Weser, but not in the EEZ (see chapter 9). Thus, in the German EEZ no area for any of the Annex II-fish species fulfils the criteria. One reason for this gap was missing scientific data for non-commercial fish species.

With regard to harbour porpoises, a large area west of the island of Sylt was selected and an essential part of it demarcated because during breeding season it is their main area of concentration in the German EEZ of the North Sea. The selection process is explained in more detail in box 1.

[6] Article 1(k) Habitats Directive 92/43/EEC.

Box 1. Site selection procedure for harbour porpoises (Annex I) in the German EEZ of the North Sea

Example of site selection for harbour porpoises (*Phocoena phocoena*) in the German EEZ of the North Sea

Regarding the criteria of Article 4.1 of the Habitats Directive, only one pSCI in the German EEZ of the North Sea was identified and in major parts delineated for harbour porpoises. The identification and demarcation was possible using the three criteria of the ad hoc meeting convened by the EC (EC 2001). There are: *"continuous or regular presence of the species"* (although subject to seasonal variations), *"good population density"* (in relation to neighbouring areas), *"high ratio of young to adults during certain periods of the year"*, and one additional: *"high proportion of sensitive behaviour"*, i.e., resting.

Selection principles:

(1) **Data collection:** harbour porpoises occurrence, distribution, and behaviour were studied by quasi-synoptical aerial transect surveys. In areas with higher occurrence of harbour porpoises additional flights with a higher density of line transects were carried out.
These data were completed and verified by long-term data sets from the observations in the Seabirds-at-Sea database (SAS), observations of local aerial data collections for environmental impact assessments, data of SCANS I, and data of porpoise detectors (POD). The last method was used successfully in the Baltic Sea only.

(2) **Species distribution maps:** harbour porpoises concentrations from May to August (reproduction time) were modelled by geo-statistical methods, based on variogram[a] analysis and ordinary kriging[b], and were visualised as distribution maps in a Geographical Information System (GIS).

(3) **Concentration areas:** concentration boundaries were identified by using density thresholds given by marine mammals expert (log 0.04 per km^2 transect).

(4) **Population size estimation:** within the pSCIs and for the whole German North Sea the population size was calculated and the proportion of harbour porpoises within given borders was estimated.

(5) **Selection and demarcation:** only one concentration site with an up to 10-times higher population density of harbour porpoises during the important reproduction period, compared to surrounding waters, was selected. Demarcation of this pSCI was mainly according to the harbour

Box 1 continued.

porpoises' density gradient. However, the distribution of the habitats sandbanks and reefs were also important delineation criteria.

Finally, the boundary lines were simplified and straightened in order to ensure simple and secure marking and identification of sites at sea.

In most of the other German offshore pSCIs, harbour porpoises occur regularly but were registered in EU data forms only as "present" because their population density does not fulfil the criteria named in Article 4.1.

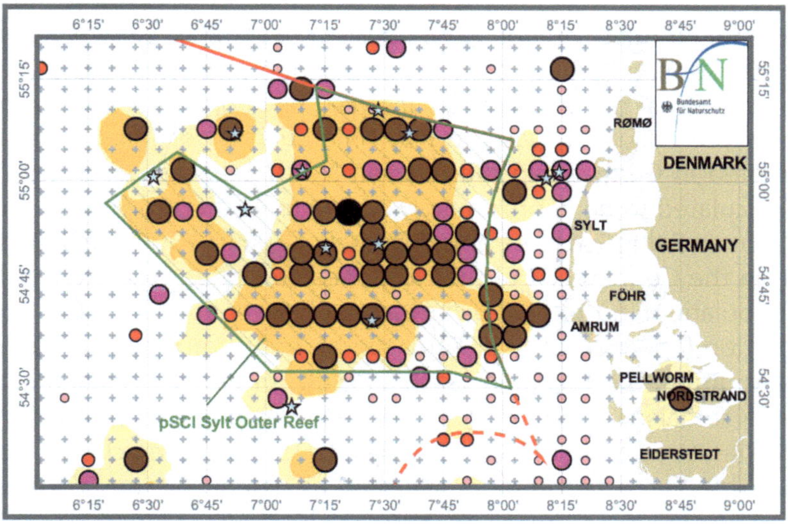

Basis: Nautical chart 2920 "Deutsche Seegrenzen, Deutsche Nordseeküste und angrenzende Gewässer"
Publisher: Federal Maritime and Hydrografic Agency

Projection: Mercator-Abbildung

Geodatic Reference System: World Geodetic System (Datum WGS 84)

0 15 30 nautical miles
(1 nm ~ 1,852 km)

Boundary of the German EEZ

Boundary of the German Territorial Waters

Depths (in meter, reference mid water level)

Country

pSCI Sylt Outer Reef

Harbour porpoises, per 3x6 min grid
(data basis: areal surveys and Seabirds at Sea Database):

- no sighting
- < 0.026 number/km
- 0.026 - 0.058 number/km
- 0.058 - 0.151 number/km
- 0.151 - 1.712 number/km
- 1.713 - 2.818 number/km
- Sightings of Harbour porpoise calves

Density distribution of Harbour porpoises based on a 2 D-Kriging algorithm; in calculation border at 6° E
(data basis: arial surveys)

Areas of high Harbour porpoise density (above threshold of log 0.04 number of individuals per km of arial survey)

Occurrence of Harbour porpoise (below threshold)

Source: www.habitatmarenatura2000.de

[a] Variogram: a measure of the variance between data as a function of distance.

[b] Kriging: a form of statistical modelling that interpolates data from a known set of sample points to a continuous surface.

Site selection criteria for Annex II species are similar to those for the Annex I habitats (stage I criteria of the Habitats Directive (Annex III)):

(a) size and density of the population of the species present on the site in relation to the population present within the national territory;
(b) degree of conservation of the features of the habitat which are important for the species concerned and for restoration possibilities;
(c) degree of isolation of the population present on the site in relation to the natural range of the species;
(d) global assessment of the value of the site for conservation of the species concerned.

In order to fulfil criterion (a) *proportion of Member State population*, the total German harbour porpoise population was estimated by quasi-synoptical aerial transect surveys analysed by geo-statistical techniques (see chapter 11). This was the first time such an estimate had been attempted in Germany. In the German North Sea, the population size was calculated to be approximately 34,000 to 40,000 individuals, and in the German Baltic Sea to be approximately 1,600 to 4,500 individuals. All pSCIs in the German EEZ of the North Sea cover approximately 40% of the population. In the pSCI *Sylt Outer Reef* (*Sylter Außenriff*), the largest proportion (approximately 35%) of the harbour porpoise population is present during the breeding season. Results of geo-statistical methods provide sufficient information to estimate the population size for the broad classes (class A: $100>p\geq15\%$, class B: $15>p\geq2\%$ and class C: $2\%>p$, with p=measured percentage) specified as default by the EC Decision 97/266.

In the easternmost waters of the German Baltic Sea, the *Odra Bank* was selected to protect a large proportion of the most threatened harbour porpoise population of Europe: the population in the Baltic proper. However, these pSCI boundaries follow mostly the habitat feature of the sandbank and the political borders.

Considering the high variability in all harbour porpoise survey data of the German waters, nonetheless, the site selection for these species appears to be sufficient according to the thresholds for the examination of Member States' proposals referred to in the biogeographical seminars (Boillot et al. 1997). These are, in principle, the same as explained for the Annex I habitats.

Many important features of the feeding and reproduction biology of harbour porpoises are still not fully understood. Therefore, it was concluded that the up-to-ten-times higher population density in *Sylt Outer Reef* area compared to the neighbouring areas gives indirect evidence for the presence of *important habitat features* for this species in that site (criterion b).

As a measure of the contribution of a population in a given site to the *genetic diversity of the species* (criterion c), the harbour porpoise populations were distinguished according to the International Whaling Commission (IWC 2000). It was concluded that the populations in the North Sea and in the western Baltic Sea are not isolated. However, the very few individuals of the eastern population (approx. 600 individuals for the entire Baltic proper) recorded for the pSCI *Odra Bank* (westernmost part of the Baltic proper) was considered to be at the margins of its distribution area (97/266/EC).

Finally, the only existing German harbour porpoise sanctuary in Territorial Waters is fully adjacent to the nominated pSCI *Sylt Outer Reef* in the EEZ of the North Sea. This was considered as an added-value factor for the last criterion (d), the *global assessment* of sites for Annex I species.

2.3 Identifying, locating, and assessing Special Protection Areas

On land and on sea, the Birds Directive (79/409/EEC, Article 4(1)) requires Member States of the European Union to undertake conservation measures concerning the habitat of: (a) bird species in danger of extinction, (b) bird species vulnerable to specific changes in their habitat, (c) bird species considered rare because of small populations or restricted local distribution, and (d) other bird species requiring particular attention due to the specific nature of their habitat, in order to ensure their survival and reproduction in the area of distribution.

According to Article 4(1) and (2), in all EEZs of the EU Member States, "the most suitable territories in number and size" must be identified as Special Protection Areas (SPAs) for both Annex I and regularly occurring migratory seabirds.

Given the above-mentioned criteria, a species priority list for the German North Sea and Baltic Sea was developed by experts (Gellermann et al. 2004, unpublished) according to the conservation status of relevant seabirds. The status was assessed by combining criteria from the AEWA (African-Eurasian Waterbird Agreement, 1999), the German Red List (Bauer et al. 2002), BirdLife and Ramsar (Heath and Evans 2000, Unselt et al. 2000) with the biogeographic population size and distribution of the seabirds.

In contrast to Annex I species (Habitats Directive), the database for the occurrence and distribution of relevant seabirds was substantially better, with more than ten years of data collected for European and German Seas and stored in the Seabirds-at-Sea database (SAS) (Garthe and

Hüppop 1996) and European Seabirds-at-Sea database (ESAS) (Reid and Camphuysen 1998). As part of the NATURA 2000 research programme, the spatial coverage was considerably further improved for all areas in the German EEZ due to additional ship and aerial surveys in the German EEZ of North Sea and Baltic Sea. For analyses, these heterogeneous data were weighed and combined (see chapter 13).

Additionally, one area in each sea was selected for further evaluation due to its designation in the list of Important Bird Areas (IBA) for the German North Sea (Skov et al. 1995) and the Baltic Sea (Skov et al. 2000).

Numeric criteria for the size of a SPA, and quality criteria as described in Annex III of the Habitats Directive, are not given by the Birds Directive. Therefore, the size of the selected SPAs in the German EEZ was determined with the proportion of the population of the relevant bird species using the percentage thresholds explained above under the Habitats Directive (Boillot et al. 1997) as guidance (Gellermann et al. 2004, unpublished). Additionally, experts were consulted in order to determine the shape and size of SPAs that would fulfil the overarching criterion of the Birds Directive to select the "most suitable territories in number and size" for the German EEZ.

In the German North Sea, the heterogeneity of the data did not allow the use of geo-statistical methods and therefore one single site was demarcated covering the wintering, resting and feeding areas of Annex I species, mainly red-throated divers and black-throated divers. In the German Baltic Sea, clusters of various seabirds, mostly marine ducks, i.e., long-tailed duck and common scoter, were much more conspicuous. Suitable sites were demarcated according to the results of geo-statistical interpolation and analysis (see chapter 13). Box 2 describes the underlying procedure.

3 Demarcation and selection of pSCI and SPA

In a final step, all relevant information on the Annex I habitats and their species, Annex II species and their habitats of the Habitats Directive, and the distribution of seabirds according to the Birds Directive were compiled in a Geographical Information System (GIS). Overlaid conservation features were then easily identified. The proportions of the included habitats and species were calculated separately for the North Sea and the Baltic Sea, since they are located in separate biogeograhic regions defined by the European Commission. These two steps were reiterated with modified site boundaries, until appropriate proportions of features according

Box 2. Site selection procedure for seabirds in the German EEZ of the Baltic Sea

Example of site selection for seabirds in the German EEZ of the Baltic Sea

The identification and selection process of Special Protection Areas (SPAs) in the German EEZ of the Baltic Sea resulted in a single large SPA of approximately 2,000 km^2 nominated to the EU Commission in May 2004. This area is defined by overlapping concentrations of more than half-a-million seabirds, primarily by the distribution and abundance of divers, Slavonian grebe, red-necked grebe, great crested grebe, black guillemot, red-breasted merganser, long-tailed duck, black scoter, velvet scoter, common scoter and common eider.

Selection principles:

(1) **Bird species selection for SPA delineation:** seabird distribution in the Baltic Sea has been studied by aerial transect surveys and from ships. From the list of 33 species to be considered for the selection of marine SPAs, eleven species of Annex I and migratory bird species were found to occur regularly in offshore areas of the German Baltic Sea and which use distinct aggregation areas.

(2) **Species distribution maps:** the bird species concentrations were modelled by geo-statistical methods, based on variogram analysis and ordinary kriging, and were visualised as distribution maps in a GIS.

(3) **Definition of important seabird concentrations:** for each of these species, concentration boundaries were identified by gradient analysis (species-specific density thresholds (logarithmic density per km^2): divers 0.1, Slavonian grebe 0.1, red-necked grebe 0.1, great crested grebe 0.1, black guillemot 0.3, red-breasted merganser 0.8, velvet scoter 1.2, common scoter 1.2, long-tailed duck 1.3 and common eider 1.3). The density value of the boundaries was used as the species- and season-specific minimum density for each relevant seabird concentration.

(4) **Converting single-species data into multi-species data:** by overlaying the concentration areas of each single species, the most important areas were identified. The respective areas and contour lines were then combined so that a set of potential conservation areas was identified.

(5) **Population size estimation:** within each concentration area, the mean density and population size for each species were calculated.

(6) **Selection and assessment of SPAs in the German Baltic Sea:** finally, the most suitable areas, in number and size, for the protection and

Box 2 continued.

conservation of species of wild birds that are listed in Annex 1 of the Directive and of regularly occurring migratory species were chosen. In the case of the German EEZ in the Baltic Sea, a single large site was sufficient to nominate adequate population numbers of the relevant Annex I and migratory bird species.

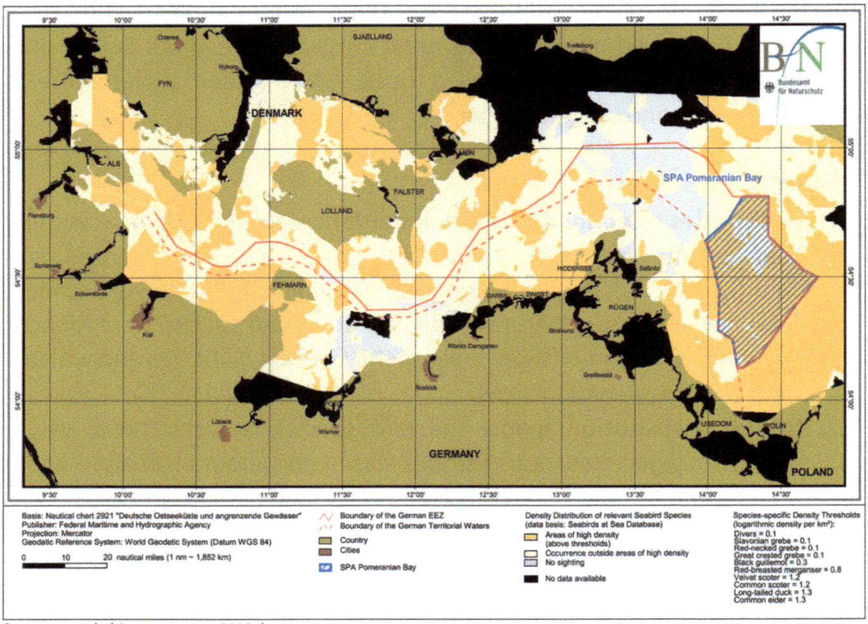

Source: www.habitatmarenatura2000.de

to the selection criteria for species and habitats were captured. Finally, the boundary lines were simplified and straightened to ensure a simpler marking and identification of the sites at sea and in shipping charts.

The ten demarcated sites were then presented and discussed with the relevant German federal ministries and the coastal States (*Länder*) and were then subjected to a public consultation process (see chapters 15 and 16). Substantial comments on the nature conservation data from this process were used to improve the data forms and to adjust the shape of one site. However, many responses that solely reflected socio-economic interests could not be taken into consideration as the Habitats Directive clearly states that for the selection of sites only conservation criteria have to be considered (Habitats Directive, Annex III stage I criteria). The so-called *Severn Decision* of the European Court of Justice affirmed this concept of the Habitats Directive (Rs. C-371/98 from 7 July 2000). Altogether, this procedure took more than one year. Finally, in May 2004, the nomination of ten NATURA 2000 sites in the German EEZ was submitted to the EU Commission in Brussels.

For sites established under the Habitats Directive, the community importance will be assessed by the Commission using the stage II criteria of Annex III after all Member States will have nominated sufficient site proposals.

4 Selected NATURA 2000 sites in the German EEZ of the North Sea and the Baltic Sea

The May 2004 nomination of ten NATURA 2000 sites within the German EEZ represents the result of about three years of preparation. Thereby, a total of 31.5% (1,040,783 hectares) of the German EEZ is covered by NATURA 2000 sites. Of these, SPAs under the Birds Directive can be protected through appropriate regulations by the German Ministry of Environment immediately after their notification to the European Commission. The appropriate steps where taken by the German Federal Environmental Ministry, and thus the two SPA sites achieved in September 2005 the legal status of a nature reserve, IUCN category IV. However, pSCIs under the Habitats Directive undergo a separate assessment process at the EU level in order to ensure the European coherence of the NATURA 2000 network. This EU level process has not yet begun.

4.1 North Sea (Atlantic Biogeographic Region)

The demarcation and important biological inventory of the three selected pSCIs and the single SPA in the German EEZ of the North Sea will be briefly described in the following.

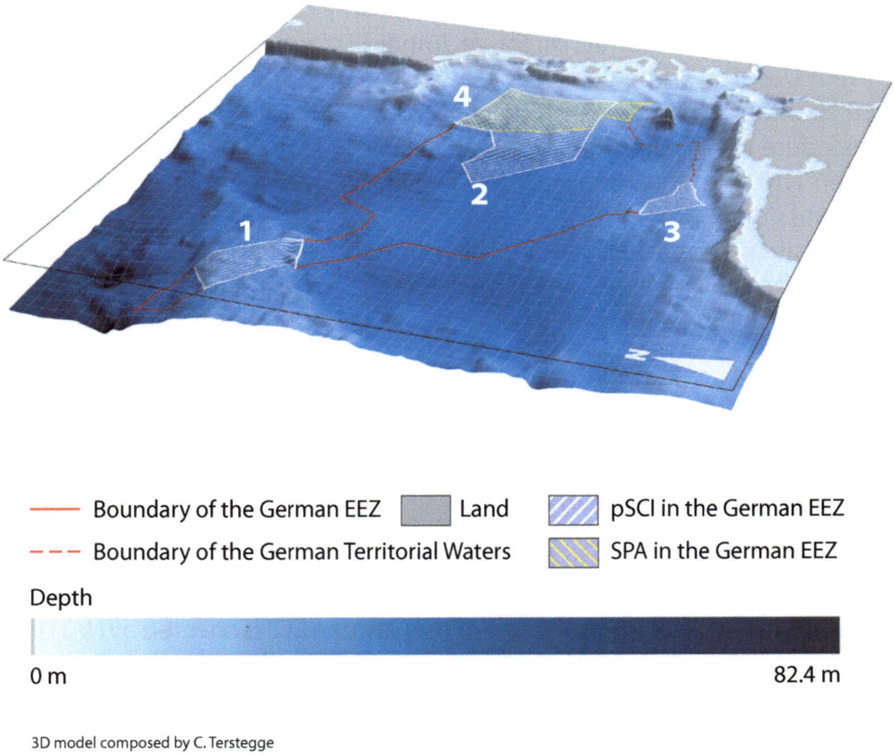

Figure 1. 3D model of the German North Sea relief and the nominated NATURA 2000 sites in the EEZ. 1 – *Doggerbank*; 2 – *Sylt Outer Reef*; 3 – *Borkum Reef Ground*; 4 – *Eastern German Bight* SPA. For illustration of the bathymetry the scale of depth was magnified. Source: Danish Hydraulic Institute (DHI), Denmark

Table 2. Key features of the nominated pSCI in the German EEZ of the North Sea

Name/ EC Code	Habitats (code)	Habitat size (km²)	Species	Population numbers
Doggerbank DE 1003-301	sandbank (1110)	1,700	Harbour porpoise (*Phocoena phocoena*)	1,558–2,508
			Harbour seal (*Phoca vitulina*)	fg
Sylt Outer Reef DE 1209-301	sandbank (1110)	app. 90	Harbour porpoise (*Phocoena phocoena*)	12,148–13,360
	reef (1170)	app. 150	Harbour seal (*Phoca vitulina*)	ep: several 1,000
			Grey seal (*Halichoerus grypus*)	ep: more than 20
			Lamprey (*Lampetra fluriatilis*)	ov
			Twaite shad (*Alosa fallax*)	ov
Borkum Reef Ground DE 2104-301	sandbank (1110)	app. 520	Harbour porpoise (*Phocoena phocoena*)	33–160
	reef (1170)	app. 23	Harbour seal (*Phoca vitulina*)	ep: several 100
			Grey seal (*Halichoerus grypus*)	ov
			Twaite shad (*Alosa fallax*)	ov

Abbreviations: **fg** feeding guest, no exact numbers; **ep** estimated population; **ov** occurrence verified, no individual numbers yet

Doggerbank

The *Doggerbank* (figure 1 and table 2) is the largest sandbank in the North Sea, dividing the sea into the ecologically distinct northern and southern regions. Most of the sandbank lies within the British continental shelf. The German portion of the sandbank deepens from the west (29 metres) to the east (40 metres), eventually extending into Danish waters as it deepens and becomes muddier. To the north and south, the sandbank borders on deep muddy slope areas (deeper than 40 metres). Both the northwestern and southeastern limits of the site correspond to the 40-metre isobath[7]. To the east and west, the pSCI site boundaries correspond to the borders of the German EEZ.

The *Doggerbank* pSCI (1,700 km²) is a sandbank representative of the open offshore sublittoral zone[8] of the central North Sea, and, at the

[7] Isobath: contour line linking regions of the same depth.

[8] Sublittoral zone: zone between the low-tide mark and the edge of the continental shelf.

Figure 2. A Thornback ray (*Raja clavata*) on the sandbank of the pSCI *Doggerbank* (photo BfN © Krause & Hübner)

same time, it is unique, with structures and ecological functions that are well preserved (figure 2). Harbour porpoises (*Phocoena phocoena*) are commonly observed with a comparatively high proportion of mother-calf groups. The presence of common seals (*Phoca vitulina*) has also been documented (Adelung et al. 2002). Due to its greater distance from land, the density of data and population surveys for this area is less intensive than that of the nearshore area of the *German Bight* (*Deutsche Bucht*).

***Sylt Outer Reef* (*Sylter Außenriff*)**

This nominated site (5,314 km^2) (figure 1 and table 2) comprises the outer grounds off Sylt and Amrum and the northeastern flank of the Pleistocene Elbe Valley. The site has been designated mainly for the following reasons:

First, it contains the largest continuously documented concentration of harbour porpoises in the entire *German Bight*, with regular mother-calf sightings and concentration "hot spots". In the latter, there has been frequent sightings of aggregations of up to 50 animals within 10 minutes which is an indication that the *Sylt Outer Reef* likely plays a role as mating habitat (see chapter 11).

Secondly, the best examples of the habitat type *reefs* (NATURA 2000 Code: 1170) in the EEZ of the German North Sea are found here (figure 3). Boulders and the typical reef species that have settled on them are, in part, overlain or surrounded by a thin layer of fine sand. This suggests that boulders and reef structures also exist below the sands, and that they are only observable on the seafloor surface where "windows" in the surface cover are present (see chapter 6).

Figure 3. Stony-reef community (with e.g., *Alcyonium* sp., *Flustra* sp., and *Pomatocerus triqueter*) in the pSCI *Sylt Outer Reef* (photo BfN © Krause & Hübner)

Thirdly, the *Amrum Bank* which lies within the site, is a typical *sandbank* (NATURA 2000 Code: 1110) of the North Friesian marine region. Ecologically, the *Amrum Bank* is characterised by high biotope and habitat diversity (according to the biotope type classification of Riecken et al. 1994), with predominantly coarse sand to gravelly slopes and fine sand areas. The habitat type is important for an infauna[9] adapted to a substrate characterised by redeposition (e.g., fast-growing zoobenthos[10]) and for long-living mussel species. This sandbank is an important stepping-stone for the coastal benthic communities in the southeastern North Sea and exhibits, above all, coarse- and fine-sand communities, such as the *Goniadella-Spisula* and *Tellina fabula* associations (see chapter 7).

This site is an important feeding area for common seals and grey seals (*Halichoerus grypus*), as well as an important marine area for the anadromous migratory Annex II-fish species, twaite shads (*Alosa fallax*) and river lampreys (*Lampetra fluviatilis*).

Borkum Reef Ground (Borkum-Riffgrund)

The *Borkum Reef Ground* (625 km²) (figure 1 and table 2) is a shallow sandbank with reef areas, lying 18 to 33 metres below sea level, located in the western part of the *German Bight*. Demarcation of the proposed site does not include the entire sandbank, but only the western part because the diversity of biocoenoses (biotopes and related ecosystems) is highest there. The great diversity of habitats of the sandbank, with its interspersed reefs, supports a high diversity of invertebrate bottom-dwellers

[9] Infauna: animals living in bottom sediments.
[10] Zoobenthos: seafloor animals.

Figure 4. Benthic community (e.g., *Metridium senile*) settling on a single stone on the sandbank of the *Borkum Reef Ground* (photo BfN © Krause & Hübner)

(epifauna[11] and infauna) among the corresponding animal communities (figure 4). Since 1998, through BfN-commissioned expeditions and investigations by Rachor (see chapter 7), 165 macrozoobenthic species[12] have been identified. Corresponding to the mosaic-like abundance of habitats, numerous small macrozoobenthos communities are present which are characterised both by a high number of macrozoobenthic species, and by numerous Red-List species (29).

In the *Western German Bight* (*Westliche Deutsche Bucht*), the *Borkum Reef Ground* is the only sandbank that extends into the deep sublittoral zone, deeper than 25 metres. In view of the primarily counter clockwise-flowing residual current in the *Western German Bight*, the *Borkum Reef Ground*, characterised by high habitat diversity, is likely an important stepping-stone for the active and passive immigration of warmwater species (e.g., *Lusitanian species*) from the southwestern North Sea and the English Channel into the water around the island of Helgoland and the Wadden Sea.

Common seals use the area as feeding ground. In addition, the site represents an important marine area for twaite shad, an anadromous migratory fish. The demarcation in the north and east conforms to the area's increasing substrate diversity. In the south and west the site is delimited by the 12-nautical mile zone and the Dutch-German EEZ border.

[11] Epifauna: animals living on the seafloor surface.
[12] Macrozoobenthic species: larger seafloor animals.

Eastern German Bight SPA (*Östliche Deutsche Bucht*)

One large single site was demarcated for the protection of bird species as a result of the analysis of long-term records in the SAS database, recent aerial surveys (MINOS 2004), and information gathered in the course of several environmental impact studies (EIA procedures) for offshore wind energy projects.

Table 3. Key features of the nominated SPA in the German EEZ of the North Sea

Name/EC Code	Species	Population numbers
Eastern German Bight DE 1011-401	Red-throated diver (*Gavia stellata*)	**w** 4,400 / **sp** 4,300
	Black-throated diver (*Gavia arctica*)	**w** 400 / **sp** 500
	Northern gannet (*Sula bassana*)	**s** 160
	Little gull (*Larus minutus*)	**w** 220
	Common gull (*Larus canus*)	**w** 11,000
	Lesser black-backed gull (*Larus fuscus*)	**b** 3,600 / **p** 2,300
	Great black-backed gull (*Larus marinus*)	**w** 800 / **a** 1,300
	Kittiwake (*Rissa tridactyla*)	**w** 800 / **b** 2,400
	Sandwich tern (*Sterna sandvicensis*)	**b** 200 / **p** 100
	Common tern (*Sterna hirundo*)	**b** 100 / **p** 100
	Arctic tern (*Sterna paradisaea*)	**b** 600 / **n** 500
	Guillemot (*Uria aalge*)	**w** 2,600 / **b** 100

Abbreviations: **w** winter; **sp** spring; **s** summer; **a** autumn; **b** breeding season; **p** post-breeding season; **n** non-breeding season

The site (3,135 km^2) comprises the *Sylt-Amrum Outer Grounds* (*Sylt-Amrumer Außengründe*) and largely corresponds to the extensions of the IBA *Eastern German Bight* (Skov et al. 1995). It is demarcated in the north by the EEZ boundary to Denmark and extends westwards to approximately 7°15' East. The eastern and southern boundaries follow the 12-nautical mile zone boundary off the North Friesian Islands and around Helgoland. The eastern boundary directly adjoins the Helgoland seabird conservation site and the Schleswig-Holstein Wadden Sea national park, both of which are classified SPAs. Water depths range from approximately 10 to 30 metres.

The SPA *Eastern German Bight* (table 3) satisfies RAMSAR criteria 5 and 6.[13] The site boundaries were determined mainly by identifying areas of

[13] Criterion 5: a wetland should be considered internationally important if it regularly supports 20,000 or more water birds. Criterion 6: A wetland should be considered internationally important if it regularly supports 1% of the individuals in a population of one species or subspecies of water bird.

high abundance over the course of the year for black-throated and red-throated divers (*Gavia stellata* and *Gavia arctica*), sandwich terns (*Sterna sandvicensis*), Arctic terns (*Sterna paradisaea*), common terns (*Sterna hirundo*), little gulls (*Larus minutus*) and common gulls (*Larus canus*). The southern boundaries of the site additionally took into account the foraging areas of the following bird species which, within German territory, only breed a little further south on Helgoland: kittiwake (*Rissa tridactyla*), guillemot (*Uria aalge*), razorbill (*Alca torda*), fulmar (*Fulmarus glacialis*), and gannet (*Sula bassana*).

4.2 Baltic Sea (Continental Biogeographic Region)

In the following, the demarcation and relevant biological inventory of the five selected pSCIs and the single SPA in the German EEZ of the Baltic Sea will be briefly described.

Table 4. Key features of the nominated NATURA 2000 sites in the German EEZ of the Baltic Sea

Name/ EC Code	**Habitats** (code)	**Habitat size** (km^2)	**Species**	**Population numbers**
Fehmarn Belt DE 1332-301	sandbank (1110)	app. 5	Harbour porpoise (*Phocoena phocoena*)	>100
			Harbour seal	fg
	reef (1170)	60	(*Phoca vitulina*)	
Kadet Trench DE 1339-301	reef (1170)	app. 25	Harbour porpoise (*Phocoena phocoena*)	>10
Adler Ground DE 1251-301	sandbank (1110)	85	Harbour porpoise (*Phocoena phocoena*)	80
			Grey seal	ov
	reef (1170)	110	(*Halichoerus grypus*)	
Western Rønne Bank DE 1251-301	reef (1170)	app. 75	Harbour porpoise (*Phocoena phocoena*)	ov
Pommeranian Bay with Odra Bank DE 1652-301	sandbank (1110)	480	Harbour porpoise (*Phocoena phocoena*)	350–850
			Sturgeon (*Acipenser sturio*)	extinct?
			Twaite shad (*Alosa fallax*)	ov

Abbreviations: **fg** feeding guest, no exact numbers; **ov** occurrence verified, no individual numbers yet

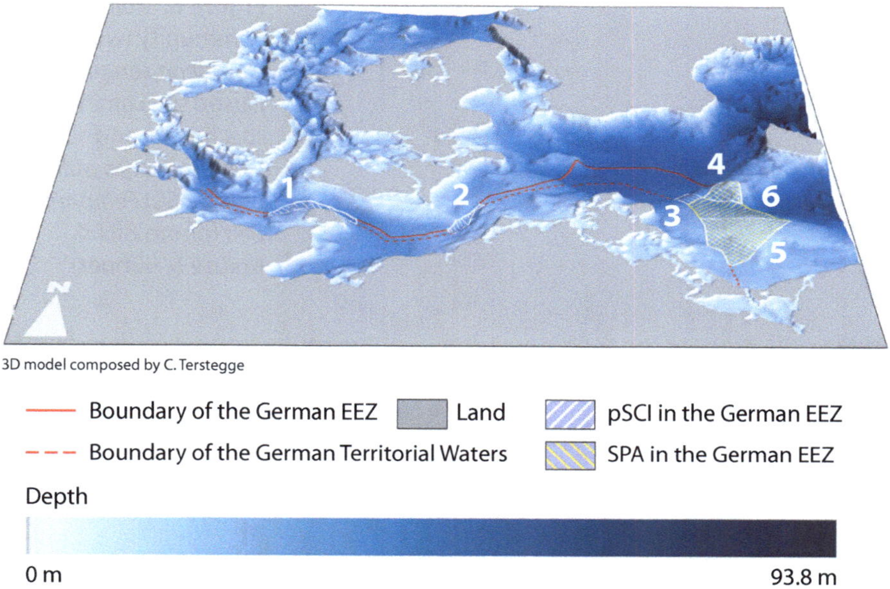

Figure 5. 3D model of the German Baltic Sea relief and the nominated NATURA 2000 sites in the EEZ. 1 – *Fehmarn Belt*; 2 – *Kadet Trench*; 3 – *Western Rønne Bank*; 4 – *Adler Ground*; 5 – *Pommeranian Bay* with *Odra Bank*; 6 *Pommeranian Bay* SPA. For illustration of the bathymetry the scale of depth was magnified. Source: Seifert et al. 2001

Fehmarn Belt (*Fehmarnbelt*)

The site (280 km^2) (figure 5 and table 4) is demarcated in the west by reefs (figure 6) and in the south by subaqueous dunes that are morphologically similar to a sublittoral dune landscape and form a subtype of sandbanks. The subaqueous dune fields are elevations of the seafloor with a same surface layer of at least 40-centimetre thickness, a slope of at least 1.0 to 0.5 degrees, and a height of one to several metres above the surrounding bottoms. Subaqueous dunes are very rare in the Baltic Sea. In German waters there is only one additional occurrence in the *Kadet Trench* known.

The *Fehmarn Belt* is a channel between the German island of Fehmarn and the Danish island Lolland in the transition area between the Belt Sea and the Baltic Sea proper. It is up to 35 metres deep and is aligned

Figure 6. Stony-reef community (with e.g., *Delesseria sanguinea*, *Flustra* sp., *Laminaria saccharina*, and *Dendrodoa grossularia*) of the pSCI *Fehmarn Belt* (photo BfN © Krause & Hübner)

in a west-to-east direction. Approximately 70% of the water exchange between the North Sea and the Baltic Sea takes place through this area. For this reason, it represents a key connection area between these two bodies of water. In terms of ecological coherence (Article 10, Habitats Directive), many marine organisms depend on an undisturbed passage through this strait during their migrations. This is especially true for the larvae of the marine species of the Baltic proper coming from areas of higher salinity in the western Baltic Sea.

Harbour porpoises occur in the proposed NATURA 2000 site and in the immediately surrounding waters of the Danish EEZ and Territorial Waters off Schleswig-Holstein and regularly swim through the *Fehmarn Belt*. The area around Fehmarn exhibits a relatively high density of harbour porpoises and a large number of animals show sensitive behavioural phases (e.g., resting phases) (see chapters 11 and 12). Common seals are regularly feeding in the pSCI. They have used the site since at least 1900 as reproduction and resting area.

Kadet Trench (*Kadetrinne*)

In terms of broad scale morphological structure, the *Kadet Trench* (100 km^2) (figure 5 and table 4) is a channel system that cuts up to 32 metres deep into the till ridge of the *Darss Sill* (*Darsser Schwelle*) between Germany and Denmark. Boulder beds occur on the slopes and on the channel floor, rising from the seafloor. These were assigned to the habitat type *reefs*. Investigations show that these are colonised by rare and threatened flora and fauna characteristic of reefs (e.g., sugar kelp, sea anemones,

Figure 7. Stony-reef community (with e.g., *Delesseria sanguinea, Laminaria saccharina,* and *Hydrozoa*) of the pSCI *Kadet Trench* (photo BfN © Krause & Hübner)

mussels) (figure 7). An especially species-rich benthic fauna (greatest numbers of Red List species in the German Baltic Sea) colonises the reefs as well as the smaller areas of heterogeneous sediments between the single boulders and boulder fields; for this reason, they are also included in the site (Gosselck et al. 1998). The *Kadet Trench* is as critical for the water exchange (approximately 70%) between the North Sea and the Baltic Sea as the *Fehmarn Belt*. Thus, it is of the same importance for the ecological coherence of the entire Baltic Sea.

Harbour porpoises occur in the *Kadet Trench* and the immediately surrounding waters (Territorial Waters off Mecklenburg-Western Pommerania) and swim regularly through the site. Although several flight surveys produced no sightings, numerous POD (Porpoise Detector) data analysed to date have documented regular occurrences of harbour porpoises (see chapter 12). This indicates that the *Kadet Trench* is likely to be important as a migratory route for this small and rare cetacean.

Adler Ground (Adlergrund)

The proposed site (234 km^2) in the area of the *Adler Ground* (figure 5 and table 4) encompasses the shallow parts of the *Rønne Bank* (*Rönnebank*) between Rügen and Bornholm, and represents the largest and shallowest underwater area in the southern Baltic Sea with the natural habitat types *reefs* and *sandbanks*. In the shallow areas (less than 10 metres), the *Adler Ground* is colonised by macroalgae, primarily *Fucus serratus, Chorda tomentosa* and *Furcellaria lumbricina* (figure 8). In the deeper boulder fields (10 to 20 metres), the common mussels (*Mytilus edulis*) dominate in

Figure 8. Stony-reef community (with e.g., *Fucus serratus* and *Mytilus edulis*) of the pSCI *Adler Ground* (photo BfN © Krause & Hübner)

the form of mussel beds (biogenic reefs). Particularly in the outer reaches of the reef, the site is dominated by sandbanks formed from glacial and marine sands. The *Adler Ground* is not normally affected during anoxia phases (lack of oxygen) in the Baltic affecting deeper waters and bottoms. Thus, it has an important function as a starting point and resource for the recolonisation of the deeper surrounding bottoms by benthic species after mass mortality events that occur regularly in the Baltic Sea due to anoxia. The northern and eastern borders follow the course of the EEZ, and are shaped to the west and south according to the sandbank structure. The border also fits partially with the boundary of the proposed SPA. Since the reefs as well as the sandbanks serve as feeding grounds for wintering bird species, the distribution of birds on the surface may be seen as an indicator of the ecological value of the site.

***Western Rønne Bank* (*Westliche Rönnebank*)**

The site (99 km^2) (figure 5 and table 4) is a largely undisturbed nearshore till ridge of the *Rønne Bank* that is interspersed to greater depths (32 metres) by boulder reefs (figure 9). In addition, the occurrence of Habitats Directive Annex II-fish species, such as the twaite shad is verified, as well as individuals of the small eastern population of the Baltic Sea harbour porpoise.

To the west, the site boundary is marked by the limit of the 12-nautical mile zone, to the north and south by the presence of reef structures.

The reef represents a link between the *Adler Ground* (above) and the reefs on the northeastern shore platform of the island of Rügen.

Figure 9. Stony-reef community (with e.g., *Mytilus edulis*) of the pSCI *Western Rønne Bank* (photo BfN © Krause & Hübner)

Pommeranian Bay with Odra Bank (Pommersche Bucht mit Oderbank)

The *Odra Bank* is the central morphological structure in the site with an area of circa 1,102 km^2. It is a typical, ideally formed large sandbank in the sense of NATURA 2000 Code 1110, and is the best example of this habitat type in the entire southern Baltic Sea. It rises significantly from the bottom of the *Pommeranian Bay* (figure 5 and table 4) and serves both as a wintering area for numerous bird species (see below), as a feeding and growth (nursery) area for young fish, and it provides excellent habitat for the bottom fauna (figure 10). Because it extends into shallower water levels, it offers a place of refuge and regeneration before and after oxygen deficiency events, as well as a starting point for the recolonisation of damaged areas (Gosselck et al. 1998). In addition, the *Odra Bank* could retrieve its importance as grazing area for the actual extinct Baltic sturgeon (*Acipenser oxyrinchus*) if its re-establishment under an on-going BfN project will be successful in the future. The site is of particular ecological importance as a nursery and spawning ground for plaice (*Pleuronectes platessa*) and especially for turbot (*Psetta maxima*).

Because of its exceptional ecological significance, the entire sandbank, its slopes and the immediate surrounding areas of the *Pommeranian Bay* are proposed as a complex protected site. The boundaries of the pSCI enclose the areas with comparatively very high concentrations (for the Baltic Sea) of harbour porpoises.

The only marine mammal in the *Odra Bank* listed in Habitats Directive Annex II is the harbour porpoise. These are probably individuals of the

Figure 10. A flounder (*Platichthys flesus*) on the sandbank of the pSCI *Odra Bank* (photo BfN © Krause & Hübner)

harbour porpoise population of the eastern Baltic Sea (Huggenberger et al. 2002), which is considered highly endangered with approximately 600 individuals only. This population is morphologically distinct from the western Baltic Sea population. There seems to be a very limited genetic exchange between the populations presumably caused by spatial separation of the reproduction areas. The *Darss Sill*, which strongly reduces the inflow of marine water masses into the Baltic Sea, is assumed to function as an obstacle that separates the populations.

Table 5. Key features of the nominated SPA in the German EEZ of the Baltic Sea

Name/EC Code	Species	Population numbers
Pommeranian Bay DE 1552-401	Red-throated diver (*Gavia stellata*)	**w** 150
	Black-throated diver (*Gavia arctica*)	**w** 150
	Red-necked grebe (*Podiceps grisegena*)	**w** 300
	Slavonian grebe (*Podiceps auritus*)	**w** 330
	Common eider (*Somateria mollisima*)	**w** 0
	Long-tailed duck (*Clangula hyemalis*)	**w** 245,000
	Common scoter (*Melanitta nigra*)	**w** 11,000 / **sp** 230,000
	Velvet scoter (*Melanitta fusca*)	**w** 55,000
	Red-breasted merganser (*Mergus serrator*)	**w** 0
	Little gull (*Larus minutus*)	**ov**
	Common tern (*Sterna hirundo*)	**ov**
	Arctic tern (*Sterna paradisaea*)	**ov**
	Black guillemot (*Cepphus grylle*)	**w** 690

Abbreviations: **w** winter; **sp** spring; **ov** occurrence verified.

***Pommeranian Bay* SPA (*Pommersche Bucht*)**

In the German EEZ of the Baltic Sea, the *Pommeranian Bay* (figure 5 and table 5) and the adjoining *Adler Ground* area are by far the most important areas for a total of several hundred thousands of individuals of long-tailed ducks, black guillemots, common scoters, velvet scoters, Slavonian and red-necked grebes, and red- and black-throated divers. The *Pommeranian Bay* is thus the most suitable SPA site in terms of both area and numbers. Within the site (2,010 km²), areas of high abundance for the first four mentioned species extend from west to east. It continues to the south far beyond the German EEZ, into the EEZs of Poland and Denmark, and into the German Territorial Waters (12 nautical miles) off Mecklenburg-Western Pommerania. Seabird concentrations are lower in the area between the *Adler Ground* and the *Odra Bank* (figure 2 and table 5). Nevertheless, this area was included in the site demarcation because, depending upon varying ice conditions in winter, it is used by wintering birds moving from south to north, and vice versa. It is thus of major ecological relevance during this period.

5 Conclusions

In May 2004, Germany successfully completed its programme for the identification, selection and assessment of habitats and species within its EEZ according to the European Nature Directives. Ten sites were nominated for NATURA 2000. Of these ten sites in the North Sea and the Baltic Sea, eight were nominated according to the Habitats Directive, encompassing approximately 2,800 km² of *sandbanks* and approximately 440 km² of *reefs*, and include important reproduction habitats of an estimated 14,000 or more harbour porpoises. Two sites were nominated according to the Birds Directive. These include, in the North Sea, the wintering and moulting grounds of more than ten thousand seabirds, mainly divers (*Gavia* spp.), terns (*Sterna* spp.) and auks; and in the Baltic Sea, those of more than 500,000 wintering seabirds, mainly divers (*Gavia* spp., *Podiceps* spp.), terns (*Sterna* spp.), and marine ducks (*Clangula* sp., *Melanitta* spp.).

Poor or nonexistent data are understandably regarded as hindrances in the establishment of offshore NATURA 2000 sites. Nonetheless, the German identification and selection programme has demonstrated that it is possible, even in the EEZ where data are particularly sparse, to collect the required EU information to sufficiently fulfil the EU data standards. These data gaps were filled in less than three years, necessitating activities such as seafloor mapping and surveying the wide-ranging harbour porpoise.

In order to meet the relevant data criteria of NATURA 2000 (Stage I Annex III of the Habitats Directive), the following two support documents were found to be helpful: Decision of the Commission 97/266/EC and the biogeographical seminars of the EU Commission.

Clearly, sound site selection must be followed by effective management if the overall conservation intent of marine NATURA 2000 sites is to be achieved. Thus, the implemented management measures for all sites should fulfil the standards demanded by the European Nature Directives.

The German EEZ nominations represent but one necessary component of an EU network. Conserving marine biodiversity in the European offshore areas demands the full participation of all maritime European Member States. It is hoped that their nominations for ecologically important offshore sites will appear in the near future.

6 References

Adelung D, Wilson RP, Liebisch N (2002) MINOS Vorhaben: Telemetrische Untersuchungen zur räumlichen und zeitlichen Nutzung des Schleswig-Holsteinischen Wattenmeeres und des angrenzenden Seegebietes der Seehunde (*Phoca vitulina*), Studie im Auftrag des BMU, Zwischenbericht 09/02 (final report in preparation) (http://www.minos-info.de)

Agardy T (1997) Marine protected areas and ocean conservation. Academic Press, San Diego, California USA

Agreement on the Conservation of African-Eurasian Migratory Waterbirds (AEWA) (1999) Report on the Conservation Status of Migratory Waterbirds in the Agreement Area. Wetlands International

Bauer H-G, Berthold P, Boye P, Knief W, Südbeck P, Witt K (2002) Rote Liste der Brutvögel Deutschlands. (German Red List of breeding birds) 3. überarbeitete Fassung, 8.5.2002. Ber. Vogelschutz 39:13–60

Birds Directive (79/409/EEC) Council Directive of 2 April 1979 on the conservation of wild birds

Boillot F, Vignault M-P, de Benito JM (1997) Process for assessing national lists of proposed sites of community interest (pSCI) at biogeographical level. Natur und Landschaft 72 (11):474–476

European Commission (1998) Applicability of Directives 92/43/EEC and 79/409/EEC to the marine environment. Letter from the European Commission to the European Wildlife Division of the United Kingdom, 20.04.1998

European Commission (2001) Interpretation note: Selection of sites for the species of Annex II *Phocoena phocoena*. Habitats Committee, Hab. 01/05

Garthe S, Hüppop O (1996) Das "Seabirds-at-Sea"-Programm. Vogelwelt 117:303–305

Gellermann M, Melter J, Schreiber M (2004) Fachliche Kriterien für die Identifizierung und Abgrenzung von marinen Besonderen Schutzgebieten (BSG) nach Art. 4 Abs. 1 und 2 der VRL bzw. Vorschlagsgebieten gemäß Art. 4 Abs. 1 der FFH-RL für die deutsche AWZ. (Criteria for identification and delineation of marine SPAs according to Art. 4.1 and 4.2 of the Birds Directive) Forschungs- und Entwicklungsvorhaben (Förderkennzeichen: 0327525) im Auftrage des Bundesministeriums für Umwelt, Naturschutz und Reaktorsicherheit

Gosselck F, Kreuzberg M, Bönsch R (1998) FuE-Vorhaben des BfN: Wissenschaftliche Grundlagen zur Ausweisung und zum Management mariner off-shore-Schutzgebiete im Bereich der Hoheitsgewässer und der Ausschließlichen Wirtschaftszone Deutschlands in der Ostsee und deren Integration in das System von Baltic Sea Protected Areas (BSPAs) (Scientific background information for the nomination and management of Baltic Sea Protected Areas (BSPAS) in the German Baltic Sea) Final report of the IfaÖ, Neu Broderstorf bei Rostock

Habitats Directive (92/43/EEC) Council Directive of 21 May 1992 on the conservation of natural habitats and of wild fauna and flora

Heath MF, Evans MI (eds) (2000) Important Bird Areas in Europe: Priority sites for conservation, vol 1. Northern Europe Cambridge, UK: BirdLife International (BirdLife Conservation Series No. 8)

Huggenberger S, Benke H, Kinze CC (2002) Geographical variations in Harbour porpoise (*Phocoena phocoena*) skulls; support for a separate non-migratory population in the Baltic proper. Ophelia 56 (1):1–12

International Whaling Commission (IWC) (2000) Report of the IWC-ASCOBANS working group on Harbour porpoises. Annex O. Journal of Cetacean Research and Management Suppl. 2:297–305

Kaiser MJ (2004) Marine protected areas: the importance of being earnest. Aquatic Conservation: Marine and Freshwater Ecosystem 14:635–638

Merck T, von Nordheim H (1996) Rote Listen und Artenlisten der Tiere und Pflanzen des deutschen Meeres- und Küstenbereichs der Ostsee. (Red lists of the threatened animals and plants of the German Baltic Sea) Schriftenreihe für Landschaftspflege und Naturschutz, Heft 44

MINOS (2004) Marine warm-blooded animals in the North and Baltic Seas: Foundations for assessment of offshore wind farms. Final report October 2004. Schleswig-Holstein Wadden Sea National Park Authority, Tönning (www.minos-info.de)

von Nordheim H, Merck T (1995) Rote Listen der Biotoptypen, Tier- und Pflanzenarten des deutschen Wattenmeer- und Nordseebereichs. (Red lists of the threatened animals and plants of the German Wadden Sea and North Sea) Schriftenreihe für Landschaftspflege und Naturschutz, Heft 44

Ray GC (2004) Reconsidering "dangerous targets" for marine protected area. Aquatic Conservation: Marine and Freshwater Ecosystem 14:211–215

Riecken U, Ries U, Ssymank A (1994) Rote Liste der gefährdeten Biotoptypen der Bundesrepublik Deutschland. (Red list of threatened biotopes of Germany). Schriftenreihe für Landschaftspflege und Naturschutz 41

Reid J, Camphuysen CJ (1998) The European Seabirds at Sea database. In: Spina S, Grattarola A (eds) Proceedings of the 1st meeting of the European Orn Union Biol Cons Fauna 102:291

Seifert T, Tauber F, Kayser B (2001) A high resolution spherical grid topography of the Baltic Sea, revised edition. Proceedings of the Baltic Sea Science Congress, Stockholm 25–29 November 2001

Skov H, Durinck J, Leopold MF, Tasker ML (1995) Important Bird Areas for seabirds in the North Sea. BirdLife International, Cambridge

Skov H, Vaitkus G, Flensted KN, Grishanov G, Kalamees A, Kondratyev A, Leivo M, Luigujõe L, Mayr C, Rasmussen JF, Raudonikis L, Scheller W, Sidlo PO, Stipniece A, Struwe-Juhl B, Welander B (2000) Inventory of coastal and marine Important Bird Areas in the Baltic Sea. BirdLife International, Cambridge

Stelzenmüller V, Maynou F, Ehrich S, Zauke G-P (2004) Spatial analysis of twaite shad, *Alosa fallax* (LACEPÈDE, 1803), in the Southern North Sea: Application of non-linear geostatistics as a tool to search for Special Areas of Conservation. International Review of Hydrobiology 89:337–351

Unselt C, Mayr C, Bauer H-G (2000) Germany. In Heath MF, Evans MI (eds) Important Bird Areas in Europe: Priority sites for conservation, vol 1: Northern Europe. Cambridge, UK: BirdLife International (BirdLife Conservation Series No. 8) 263–340

Part III: Identification and assessments of habitats

Chapter 5

Identification of submarine banks in the North Sea and the Baltic Sea with the aid of TIN modelling

Andreas Klein

ARGUMENT, Kiel

Abstract

This research project was conducted by the consultant ARGUMENT to develop an intersubjectively-revisable method[1] for the identification and demarcation of sublittoral banks[2] in the North Sea and the Baltic Sea by means of morphometric methods. This was done with the aid of TIN modelling and GIS-supported three-dimensional analysis of bathymetric data.

Three main steps were taken:

- collecting morphometric data and examining them vis-à-vis the aim stated above;
- developing a useful morphometric definition of marine banks which can be applied to the available set of data;
- determining and recording the distinguishing features of subtypes of marine banks in the North Sea and the Baltic Sea.

The following results were achieved:

A morphometric definition for marine banks could be developed and accordingly, identifications and demarcations of banks were possible, as long as a small window of opportunity was left open for the scientist's

[1] Intersubjectively-revisable method: all individuals who follow this method come to the same result.

[2] Sublittoral: zone between the low-tide mark and the edge of the continental shelf (here about 200m depth).

own decisions. In this regard, the density of data was a decisive criterion for the identification and demarcation of marine banks. As results, 20 marine banks were identified in the North Sea, and 63 in the Baltic Sea.

1 Introduction

Since the new German Federal Nature Conservation Act was released in 2002, the network of protected areas with the title NATURA 2000 had to be extended into the German offshore marine areas of the Exclusive Economic Zone (EEZ). For that reason, the German Federal Agency for Nature Conservation (BfN) had to identify proposed Sites of Community Importance (pSCI) in the EEZ of the North Sea and the Baltic Sea (see chapter 4). Sublittoral sandbank habitats were one interest feature (see chapter 3). Since sandbanks can be regarded as elevations of the seabed, a detailed determination of submarine banks that is suitable for evaluation needed to be developed at the request of BfN. A further request to ARGUMENT was to locate and demarcate sublittoral banks in the North Sea and the Baltic Sea on the basis of best available bathymetric data[3]. This was done with the aid of TIN-modelling and GIS-supported 3D analysis as shown in this chapter.

It was not the purpose of the research project to determine and identify sites which would represent the NATURA 2000 habitat *sandbank*. This could only be done by overlay of bank boundaries with additional information layers on sediment distribution and biology, which was done by BfN itself (see chapter 3).

2 Procedure

Three main steps were taken during the project:

- collection of data and examining them with regard to the conditions stated above;
- development of a definition of marine banks which can be applied to the available set of data;
- determination and recording of distinguishing features for all subtypes of marine banks occurring in the North Sea and the Baltic Sea.

[3] Bathymetric data: observed depths.

2.1 Collecting and analysing the data

2.1.1 Data supplied by BSH

Surveys of the Federal Maritime and Hydrographic Agency (BSH) from 1984 to 2001 provided basic digital information on depths in many marine sub-areas in German waters (Bundesamt für Seeschifffahrt und Hydrographie 2002). However, this stock of bathymetric data did not cover the whole area that was to be examined. Hence, for the North Sea, gaps were filled by digitising analogous sea maps. For the Baltic Sea, data from former research projects were available to complement necessary information.

Since BSH digitised its soundings mainly for navigational purposes, only the highest representative points on the graph lines for certain areas had been taken. Therefore sea charts or new maps produced with these data tended to reflect shallower depths than they were in reality. According to BSH there was a deviation of up to 30 centimetres in the depth readings. The precision of the geographical location of the measured points was given in the individual data sets, and was on average below 5 metres. The number of digitised points from the graph of the echo sounder was usually much higher than the distance between the single back and forth turns of the ship (approximately 10:1). Hence, especially for the North Sea, there was no evenly distributed set of data available on which further research could be based.

Soundings for certain years had not been checked by BSH before the submission of the data set. During our analysis, several errors were found (e.g., the depth at a certain point differed extremely from the surrounding area). These errors were removed from the data set. In addition, the data reflected several structures which were obviously man-made, for example in the *Eckernförder Fjord* (*Eckernförder Bucht*) or in front of the *Brodtener Stone Reef* (*Brodtener Steinriff*) (Baltic Sea). It was not possible to examine whether these structures actually exist. However, it was not the aim of the project to verify and correct the complete data set provided by BSH; this was done only for those data relevant for the identification of sublittoral banks. Missing data for the Baltic Sea were supplemented by grid data (100x100 metres) calculated by BSH for an earlier project. These data did not give real soundings but the average depth of the area covered by the grid-cell derived from the volume of water. This meant that these data could not be directly compared to the available soundings.

---

[4] Geo-referencing: digitally attaching geographical position data to the scanned map.

2.1.2 Supplementation of missing data

The missing information concerning the North Sea was supplemented by data from analogous sea maps (1:100,000 sheet "1507 Deutsche Bucht Süd"). For this purpose the maps were scanned, geo-referenced[4] and digitised (approximately 4,500 points). The data were checked by comparing and adjusting them to digital information already present at that time. The location of the points to be measured was less precise than the digital data because the points measured were not directly marked on the maps but indirectly, as the centre of the first digit of the depth indicated on the map. In addition, the preceding steps were possible sources of mistakes. In the area to be examined, the ship took echo-soundings in straight lines that were one kilometre apart. This resulted in an approximate deviation of up to 50 metres with regard to the exact position of the point being measured. However, the precision of the data was sufficient for the given scope of the project.

2.2 Definition and description of marine banks

A description of the morphometric attributes suitable to serve as a model had to be developed. The starting point for this description was the definition for the NATURA 2000 habitat *sandbanks which are slightly covered by seawater all the time*, according to the Habitats Directive and BfN (see chapter 3). These definitions contain morphological as well as geological and biological or ecological components. Consequently, the model to be developed had to meet the following requirements:

- A model should reflect the real situation as exactly as possible; that is, the description in question should correspond with the definitions of the EU as given in the Interpretation Manual (European Commission 1999), with the interpretations of BfN (Ssymank et al. 1998, Balzer et al. 2002) and with the notes of experts in that field of science (see also chapter 3).
- A model should be as simple and as clear as possible. This requirement was based on the necessity to apply the model to further projects and to making it comprehensible not only to experts in that scientific field but also to the broader public, administrative bodies and the political sector. That is, the model had to be suitable for intersubjective evaluation.
- A model had to be unequivocal. The definition should not include structures other than those to be examined.
- The model had to be operative. This general requirement was especially important for this project since the definition had to be applied to a

data set with limitations.

The discussion on this subject among experts led to three basic premises concerning the identification of submarine banks:

- Submarine banks are below sea level.
- They are mainly surrounded by slopes.
- They are mainly surrounded by deeper water.

The description of marine banks also had to include rules for demarcation. According to the premises outlined above, borders for the transition from the slopes of the bank to the surrounding plains had to be distinguished.

The possibilities of defining submarine banks with the help of such a model were not, however, unlimited. Firstly, there was a limit to how small – but not to how big – a bank to be defined could be: it had to be big enough to be drawn on a map with a scale of 1:375,000. Secondly, the possibilities were restricted by the spatial density of the data set which had to be high enough to show that certain structures are banks in the sense of the definition.

Notwithstanding such limitations, an operative description of marine banks was developed. Basically two different approaches seemed to be possible. On the one hand, parameters for the definition could be deduced directly from the data set as such, without the help of information from research in geographical or other fields. On the other hand, the definition could be derived from the analysis of such structures which so far have been accepted by experts as *banks*. The present research tried to combine the two basic approaches by supporting and illustrating the purely morphometrical description with the help of test areas.

Marine banks were thus described as follows:

- Submarine banks are permanently submerged.
- They can be distinguished as independent elevations of the seabed.
- Their boundaries are generally marked by slopes of more than 0.5°. However, if the density of data is low (North Sea), slopes of up to 0.1° can also be included.
- Boundaries are generally drawn at the transition from the slopes of the bank into the surrounding plains.
- In more level areas they are marked by the straight line between the ends of the slopes as defined above.
- The line marking the slope area should be at least 3 times longer than the straight line (see figure 3).
- The banks that the model accounts for must be bigger than 1 km^2.

2.3. Determination of marine banks with the aid of TIN modelling

For the identification and demarcation of sublittoral banks, structures which correspond to the simplified model were searched for in the North Sea and the Baltic Sea. This was done by using the stock of data compiled as described above, in order to produce a TIN model of the seafloor topography of the North Sea and the Baltic Sea. Such a Triangular Irregular Network (TIN, figure 1) could represent the connections between all individual bathymetric measurements represented by x, y and z-coordinates (Bill 1999, Zölitz-Möller et al. 1997) and was generated by triangulating all available depth points.

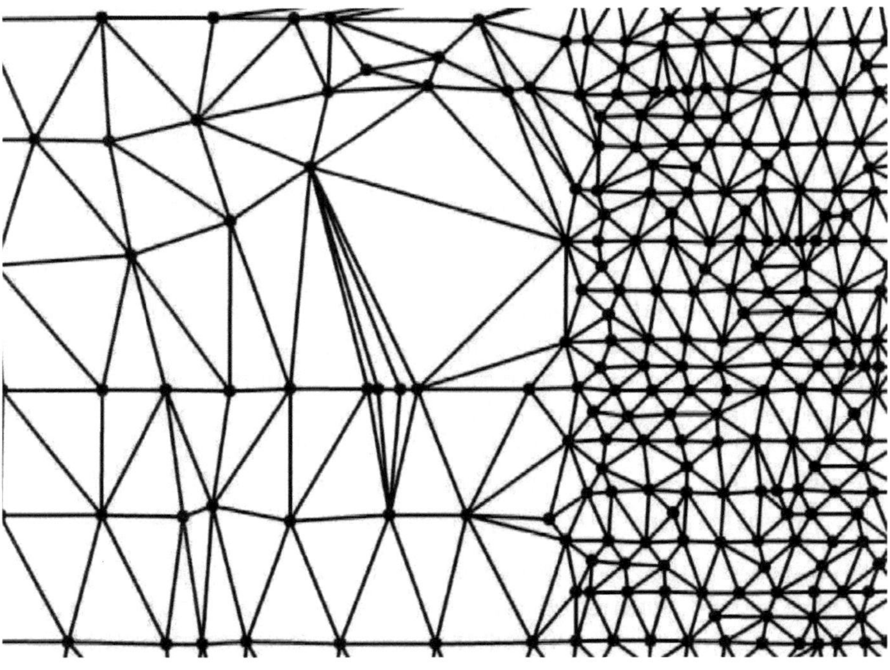

Figure 1. A TIN represents the connections between measurements represented by *x*, *y*, and *z*-coordinates

Certain attributes such as inclination, exposition, and average depth of the (triangular) areas produced by the TIN could be calculated and recalled with the help of a Geographical Information System (GIS). The seafloor of the North Sea and the Baltic Sea were analysed in this manner in order to receive an initial of submarine banks. For this purpose, the steepness of the banks' slopes was of particular importance (figure 2).

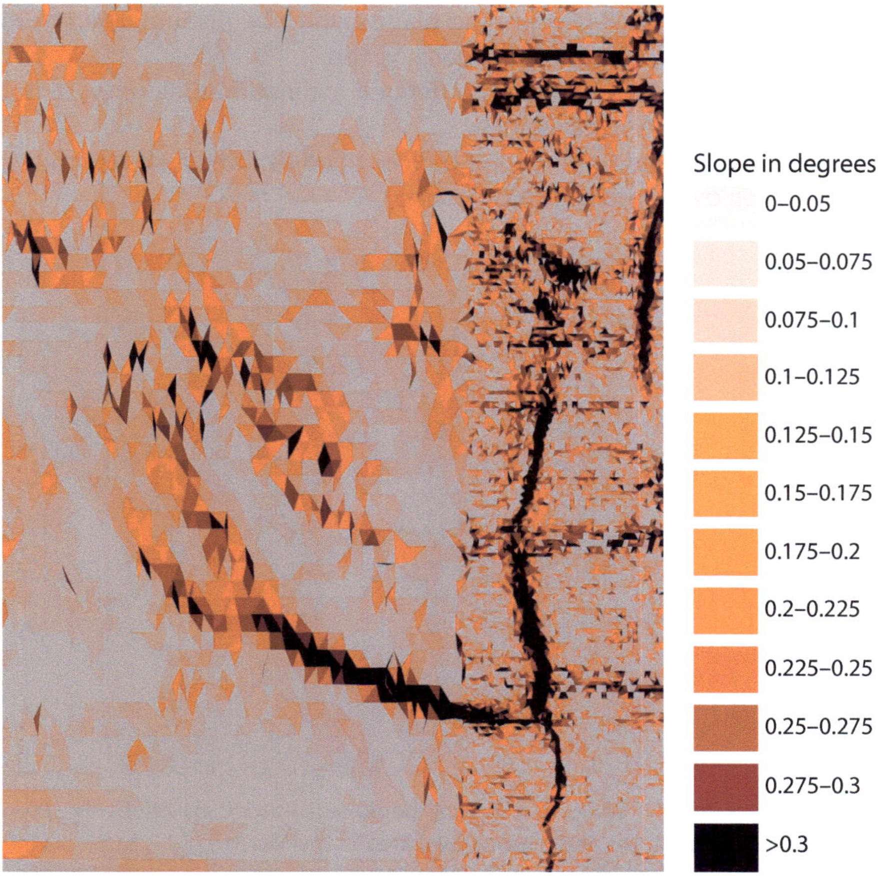

Figure 2. Inclinations calculated for the area around *Amrum Bank* (North Sea)

The *Amrum Bank* (figures 2 and 3) is, according to the above determination of sublittoral banks, an ideal example of a typical bank. It is surrounded by three edges with slopes of at least 0.5°; it levels off more smoothly only in its northern part.

However, given that the complexity of the natural conditions was only partly reflected in the available data set, not all banks could be identified by the model. Consequently, automatic devices could not always be used immediately. The following decisions for which a computer program is not suitable had to be made first by the analysing scientist:

- Inclined areas above the defined limit which represented the slopes of the bank into the surrounding plains had to be distinguished from areas with a structured relief surface on top.
- Slopes did not always reach the defined inclination fully or were not represented in the data as such.
- Marine banks could not always be identified as individual structures; they often overlapped each other.

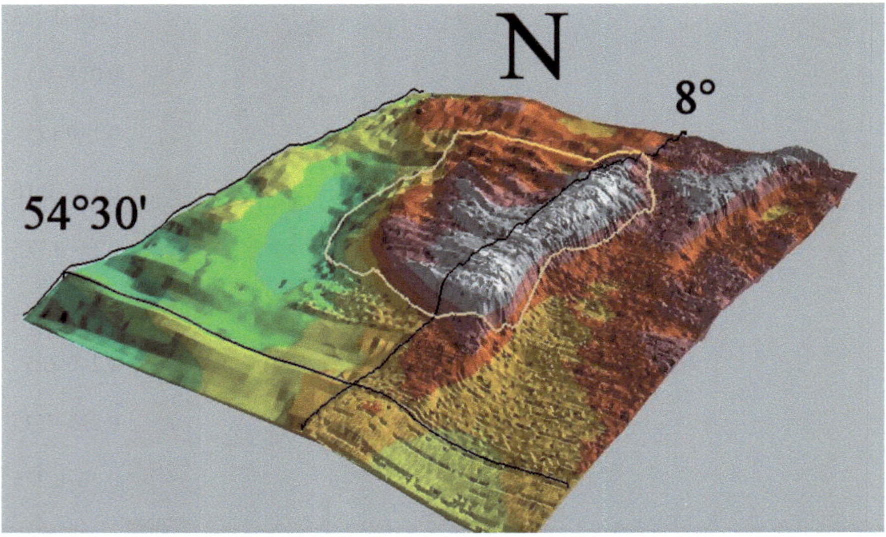

Figure 3. The *Amrum Bank* as a typical bank, surrounded by relatively steep slopes in three directions and a more level area in the North

The marine banks were mapped and demarcated according to the definition laid out in the previous section. Four different types of demarcation were recorded:

- The slopes demarcating definite transitions to the surrounding plain were determined as border lines.
- In more level areas, connecting lines between the ends of the slopes were seen as border lines.
- The boundaries of the EEZ were marked separately as boundaries of marine banks.
- In one case (*Odra Bank* (*Oderbank*) – Baltic Sea) a depth-line was taken as border line for one of its sides against the surrounding plain.

The bank polygons were transferred into a Geographical Information System (GIS) and analysed. With the help of the GIS the following data were added: the area, the minimum and maximum depth, as well as the relative height derived from these two items, the length of the individual boundaries, and the relations between them.

3 Results

Twenty marine banks were identified in the North Sea, and sixty-three in the Baltic Sea. The comparison of the marine banks in the North Sea and the Baltic Sea clearly showed the following differences:

North Sea (see table 1 and figure 4)

Fewer banks could be detected in the North Sea than in the Baltic Sea. This was partly due to the different geological history of the two seas and their different seabed morphology. Further, the density of data in the EEZ of the North Sea was distinctly smaller than for the Baltic Sea. Consequently, structures found on sea charts such as the *White Bank* (*Weiße Bank*), the *Sylt Outer Reef* (*Sylter Außenriff*), or the *Northern Shell Ground* (*Nord-Schillgrund*) were not regarded as banks in the sense of the definition given above.

Table 1. Identified banks of the North Sea
(see figure 4; names for banks in the EEZ are in bold)

	Name	Area (km^2)	Max depth (m)	Min height (m)	Height (m)
1	***Doggerbank***	**2,334.355604**	**-69.3**	**-29.4**	**39.9**
2	***Borkum Reef Ground***	**1,040.627805**	**-33.5**	**-4.9**	**28.6**
3	*Ballonplate*	42.907489	-17.7	-0.5	17.2
4	*Geldsackplate*	56.259008	-16.5	0.0	16.5
5	*Brauerplate*	101.413278	-19.9	0.0	19.9
6	*Nordwestgründe*	16.305814	-14.7	0.0	14.7
7	*Robbenplate*	17.984205	-22.1	0.0	22.1
8	*Norderriff*	14.939229	-21.3	0.0	21.3
9	*Roter Sand*	93.295676	-24.8	0.0	24.8
10	*Nordergrund*	145.694849	-22.4	0.0	22.4
11	*Scharhörnriff*	88.886527	-26.2	0.0	26.2
12	*Grosser Vogelsand*	128.124277	-24.6	0.0	24.6
13	*Buschsand*	123.957982	-15.8	0.0	15.8
14	*Blauortsand*	21.203685	-13.2	0.0	13.2
15	*Steingrund*	135.583559	-33.0	0.0	33.0
16	*Rochelsteert*	60.194308	-11.8	0.0	11.8
17	***Amrum Bank***	**151.060588**	**-22.4**	**0.0**	**22.4**
18	*Kniepsand*	60.478059	-16.9	0.0	16.9
19	*Theeknobs*	30.383703	-33.1	0.0	33.1
20	*Salzsand*	11.484882	-28.3	-0.3	28.0

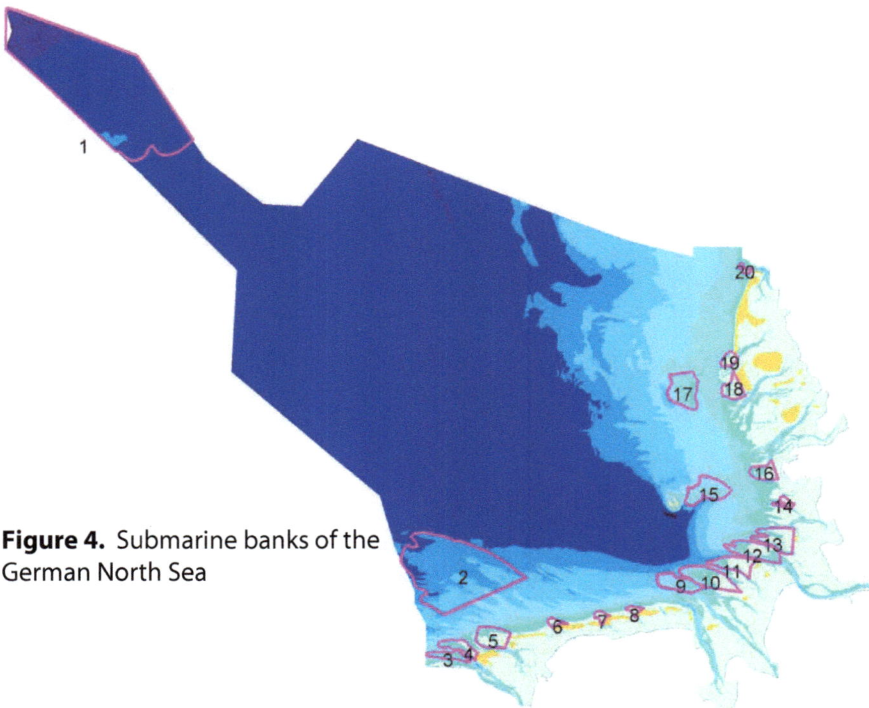

Figure 4. Submarine banks of the German North Sea

Figure 5. Submarine banks of the German Baltic Sea

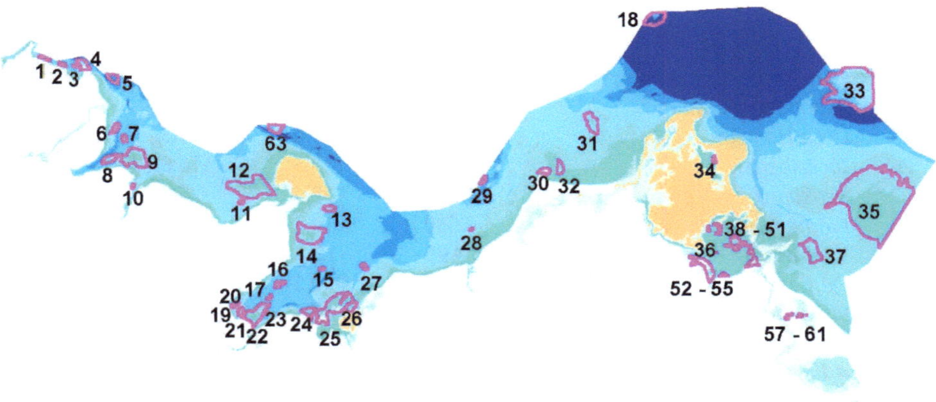

Table 2. Identified banks of the Baltic Sea
(see figure 5; names for banks in the EEZ are in bold)

	Name	Area (km²)	Max depth (m)	Max height (m)	Height (m)
1	*Langballigbank*	2.597	-7.4	-22.9	15.5
2	*Neukirchengrund*	1.989	-3.3	-25.9	22.6
3	*Geltinger Bucht*	0.587	-13.1	-24.5	11.4
4	*Kalkgrund*	11.196	-0.5	-30.7	30.2
5	*Bredgrund*	8.937	-10.1	-31.3	21.2
6	*Schwedeneck 1*	6.314	-11.9	-24.5	12.6
7	*Schwedeneck 2*	3.931	-14.9	-24.7	9.8
8	*Mittelgrund*	10.231	-6.3	-27.4	21.1
9	*Stoller Grund*	44.320	-6.6	-23.5	16.9
10	*Strander Grasberg*	1.257	-4.0	-17.3	13.3
11	*Putlosbank*	2.197	-12.3	-16.7	4.4
12	*Fehmarn Belt*	80.115	-5.6	-16.9	11.3
13	*Staberhukbank*	6.502	-14.5	-21.0	6.5
14	*Sagasbank*	51.985	-1.0	-21.9	20.9
15	*Dahmeshöveder Grund*	1.106	-19.8	-22.6	2.8
16	*Walkyriengrund*	4.999	-7.2	-25.7	18.5
17	*Pelzerhakengrund*	1.266	-19.2	-23.2	4.0
18	***Kriegers Flak***	**26.949**	**-11.0**	**-44.6**	**33.6**
19	*Haffkrugbank*	1.703	-11.0	-44.6	33.6
20	*Scharbeutzbank*	1.462	-8.8	-16.9	8.1
21	*Timmendorfer Bank*	4.786	-9.3	-20.8	11.5
22	*Brodtener Steinriff*	33.407	-1.7	-25.2	23.5
23	*Travegrund*	0.693	-16.8	-23.1	6.3

Table 2. continued

	Name	Area (km²)	Max depth (m)	Max height (m)	Height (m)
24	*Boltenhagener Grund*	10.666	-3.9	-23.1	19.2
25	*Lieps/Hannibal*	66.778	-0.6	-21.3	20.7
26	*Jöckelberg*	9.680	-2.6	-16.8	14.2
27	*Rerikbank*	3.077	-12.9	-25.0	12.1
28	*Warnemündebank*	0.630	-14.2	-20.0	5.8
29	***Yder Knob***	**3.499**	**-12.3**	**-30.1**	**17.8**
30	*Darßer Ort*	7.038	-0.5	-14.6	14.1
31	*Platagenetgrund*	24.654	-7.1	-16.1	9.0
32	*Prerow Bank*	9.712	-4.2	-9.0	4.8
33	***Adler Ground***	**209.436**	**-6.2**	**-32.7**	**26.5**
34	*Jasmundbank*	2.326	-1.7	-7.0	5.3
35	***Odra Bank***	**480.701**	**-5.9**	**-15.1**	**9.2**
36	*Grabowbank*	0.129	-4.4	-15.3	10.9
37	*Usedombank*	35.944	-4.4	-15.3	10.9
38	*Tempelberggrund*	0.671	-3.5	-7.8	4.3
39	*Schabernakbank*	0.881	-3.2	-7.7	4.5
40	*Trendelriff*	0.355	-3.7	-8.3	4.6
41	*Vilmgrund 2*	1.225	-2.9	-8.2	5.3
42	*Vilmgrund*	3.180	-2.6	-8.9	6.3
43	*Vilmgrund 3*	0.330	-4.6	-8.3	3.7
44	*Orientgrund*	0.471	-3.8	-8.9	5.1
45	*Gross Stubber*	10.772	-2.8	-10.8	8.0
46	*Gross Stubber Süd*	3.802	-0.4	-10.0	9.6
47	*Rugiagrund*	0.982	-3.6	-7.9	4.3
48	*Kleinstubber*	0.297	-2.1	-9.5	7.4
49	*Böttchergrund*	0.590	-3.6	-9.3	5.7
50	*Schumachergrund*	3.020	-2.9	-9.8	6.9
51	*Freesendorfer Haken*	8.511	-0.5	-9.0	8.5
52	*Gahlkower Haken*	2.533	-1.5	-7.3	5.8
53	*Kooser Haken*	23.852	-0.5	-7.0	6.5
54	*Reinberggrund*	1.328	-1.5	-4.6	3.1
55	*Gristower Bank*	2.481	-0.5	-4.8	4.3
56	*Wussowbank*	0.169	-2.2	-4.3	2.1
57	*Trockenort*	1.435	-1.1	-4.8	3.7
58	*Grussowbank*	0.195	-1.1	-4.8	3.7
59	*Hohe Schaar 3*	0.244	-1.7	-4.0	2.3
60	*Hohe Schaar 2*	0.360	-1.3	-3.7	2.4
61	*Hohe Schaar 4*	0.131	-1.7	-3.5	1.8
62	*Hohe Schaar*	1.093	-0.6	-4.6	4.0
63	***Fehmarn Belt Bank***	**12.499**	**-12.2**	**-28.5**	**16.3**

Baltic Sea (see table 2 and figure 5)

In the Baltic Sea, smaller structures were taken into account although they did not meet the requirements of the definition with respect to their size as individual structures. However, because of their typical distinctive feature as parts of other structures, they were not removed from the data set.

So-called "banks" which surrounded an island, for example the Isle of Vilm (Baltic Sea) were not regarded as banks because according to the definition, a bank must be mainly surrounded by slopes and deeper water and the whole area must be below sea level. For the same reason, only those parts of peninsulas which are extensions into the sea are banks by definition.

Without going into the details of each reason, it can be claimed that mapping the marine banks in the way described above provides the analyst with various types of banks:

- Typical banks are mainly isolated elevations of the seabed within a rather shallow area. They are totally surrounded by relatively steep slopes, as for example the *Walkyrien Ground* (*Walkyriengrund*), *Lieps* or *Hannibal*.
- "Attached" banks are the extensions from land into the sea. Towards the seabed they are demarcated by slopes, whereas they level-off at the borderline to the mainland.
- "Societies" of mainly smaller banks occur in some areas, as for example the *Lübeck Bight* (*Lübecker Bucht*) or the *Greifswald Lagoon* (*Greifswalder Bodden*).

Given that mapping marine banks in the North Sea and the Baltic Sea is the fundamental result of this research, the following points can be regarded as additional, more general results.

A morphometric description of marine banks and their identifications were possible as long as a small window of opportunity was left open for own decisions of the analyst. Further, the data collected and analysed in this research provided the basis for further morphological research in this field and further realisation of the potentially improved definition of marine banks.

The density of data was a decisive criterion for the identification and determination of marine banks. The spatial density of readings concerning structures to be regarded by experts as "banks" has to be around 100 metres. As a result, some banks in the North Sea might not have been identified.

References

Balzer S, Boedeker D, Hauke U (2002) Interpretation, Abgrenzung und Erfassung der marinen und Küstenlebensraumtypen nach Anhang I der FFH-Richtlinie in Deutschland – Natur und Landschaft 77 (1):20–28

Bundesamt für Naturschutz (BFN) (2002) Auswahl der NATURA 2000 Meeresschutzgebiete. Ergebnisbericht. 2. Statusseminar im Rahmen der naturschutzorientierten AWZ-Forschung.
(http://www.bfn.de/marinehabitate/en/downloads/berichte/Statusseminar_AWZ-Forschung_2002.zip)

Bill R (1999) Grundlagen der Geo-Informationssysteme, Band 1 und 2, Heidelberg

Bundesamt für Seeschifffahrt und Hydrographie (2002) Tiefendaten unterschiedlicher Räume und Jahrgänge von Nord- und Ostsee

European Commission (1999) Interpretation Manual of European Union Habitats. EUR 15/2. DG Env., Nature protection, coastal zones and tourism

Ssymank A, Hauke U, Rückriem C, Schröder E (1998) Das europäische Schutzgebietssystem NATURA 2000. Landwirtschaftsverlag, Münster-Hiltrup

Zölitz-Möller R, Heinrich U, Nachbar M (1997) The use of geographical information and the ecosystem approach in environmental planning. In: S Hodgson, M Rumor, JJ Harts (eds) Geographical Information 1997. From Research to Application through Cooperation. vol. 1, IOS-Press. Amsterdam, Berlin pp 507–516

Chapter 6

Identification of submarine hard-bottom substrates in the German North Sea and Baltic Sea EEZ with high-resolution acoustic seafloor imaging

Markus Diesing and Klaus Schwarzer

Institute of Geosciences, Christian-Albrechts-University, Kiel

Abstract

Submarine hard bottoms (e.g., boulders, outcropping strata) are of particular ecological importance. They were investigated in the Exclusive Economic Zones (EEZ) of the German North Sea and the Baltic Sea, using high-resolution seafloor imaging techniques (i.e., sidescan sonar and multibeam echosounder). Examples are shown from the research areas *Sylt Outer Reef* (*Sylter Außenriff*; North Sea), *Kadet Trench* (*Kadetrinne*), and *Adler Ground* (*Adlergrund*) (both in the Baltic Sea). There exist distinct differences between the two continental shelf seas regarding the distribution of boulders and the density (percent coverage) of boulders per unit seafloor. The observed differences are attributed to (a) different geological evolution of the seafloor, and (b) different forcing by waves, tides and currents, which are responsible for the redistribution of sediments.

1 Introduction

Submarine hard bottoms such as boulders or outcropping strata generally support a diversity of marine plant and animal species and are therefore of great ecological importance. Detailed knowledge of the locations of submarine hard bottoms is a prerequisite for their further investigation, protection, and management. Unfortunately, published maps of the seabed geology of the German North Sea and Baltic Sea shelf areas lack the resolution necessary for detailed habitat analysis. The reason for this is that such maps were mainly produced based on surface samples (e.g., Figge 1981, Tauber and Lemke 1995, Tauber et al. 1999), which then introduced errors, whether numerical or manual, during the process of interpolation. Furthermore, boulders are probably underrepresented owing to the fact that the sampling instruments used are mainly constructed to sample mud, sand, and gravel.

These problems can be addressed by the application of acoustic seafloor imaging techniques such as high-resolution sidescan sonar or multibeam bathymetric technology (e.g., Cochrane and Lafferty 2002, Kostylev et al. 2001, Ojeda et al. 2004). A large body of sidescan sonar data from the German North Sea and Baltic Sea already exists (e.g., Niedermeier-Lange and Werner 1988, Schulz and Tauchgruppe Kiel 1983, Tahrir 1984, Werner et al. 1974, 1976, Werner 2004, Winn and Werner 1984, Winn et al. 1982), but results are spatially restricted. Moreover, data were gathered as analogue paper records and with limited positioning accuracy. Thus, we collected new data on sediment distribution patterns and seafloor topography in pre-defined research areas in the German North Sea and Baltic Sea EEZ (between 12 and 200 nautical miles from the coast) in order to provide a basis to outline the occurrence of submarine boulders.

2 Regional setting

During the last glacial maximum (ca. 21,000 years ago), the global sea level was ca. 125 metres lower than present levels (e.g., Fleming et al. 1998), exposing the world's continental shelves to subaerial processes. The subsequent flooding ("marine transgression") of the continental shelves during deglaciation took place with high rates until ca. 7,000 years ago, followed by a slower rate. Due to this relatively short time span, the continental shelves of the world's oceans presently exhibit relict features (Emery 1968) which are inherited from this subaerial exposure and modified by marine transgression. Typical examples in formerly glaciated regions are moraines deposited by glaciers and river valleys, which are sometimes deeply incised due to the formerly low-lying sea level. The veneer of marine sediments, which have been deposited since the flooding of the shelves, is still relatively thin (in metre-scale) or even absent, except in areas of high sediment input (e.g., river deltas and former river valleys).

According to the map of Figge (1981), large areas of the seafloor of the German North Sea are covered with sand of varying grain-size composition. The thickness of the marine sand veneer generally amounts to less than 2 metres (Figge 1981). Thicker marine deposits are restricted to the NW–SE-trending Pleistocene Elbe valley, where silty fine sands accumulate (Figge 1980), and to the shoreface of the *German Bight* (*Deutsche Bucht*) in water depths down to ca. 10 metres below mean sea level (Zeiler et al. 2000).

The Baltic Sea can be regarded as an epicontinental shelf sea. The sediment distribution in the southwestern Baltic Sea is much more heterogeneous compared to the North Sea. Although the sediment distribution is strongly affected by underlying geologic framework, a depth-dependent overall zonation of seabed sediments can be found (Seibold et al. 1971). Coarse-grained lag deposits (coarse sand, gravel and boulders) form a thin veneer (of a few decimetres) covering morainal deposits, mainly in water depths of 5 to 15 metres along the coasts and on submarine sills. Those sediments are the result of abrasion of morainal material. Fine material up to sand size is removed by waves and currents, leaving the coarser constituents behind. Such lag deposit areas are surrounded by well-sorted fine-to-medium sands. Apart from the immediate proximity of the coast, this sand veneer is relatively thin, for example, 0.5 to 2 metres in the *Kiel Bight* (*Kieler Bucht*) (Seibold et al. 1971). Significant amounts of marine sediments are found in deeper basins and channels of the Baltic Sea, where fine-grained, organic-rich mud accumulates.

In the following, data from three sites are presented. From the North Sea, we show results of an area 70 to 100 km west of Sylt Island in water depths of 25 to 40 metres (*Sylt Outer Reef*, figure 1). From the Baltic Sea, results are presented from *Adler Ground*, located between Rügen and Bornholm Island in water depths of 5 to 25 metres, and *Kadet Trench*, where the research area is located between Fischland-Darss-Peninsula and Falster Island in waters of 10 to 30 metres deep (figure 2).

3 Methods

We employed a towed Klein 595 (Klein Associates Inc.) dual-frequency (100 and 384 kHz) sidescan sonar system (see figure 3) and a hull-mounted Seabeam 1185 (Elac Nautik) 180 kHz multibeam swath-bathymetry system, which also collects co-registered acoustic backscatter. The recorded spatial patterns of backscatter intensity were interpreted in terms of seafloor relief and sediment type. Validation of the data was achieved by grain-size analysis of seafloor sediments and by underwater video surveying. Additionally, single-beam echosounder data were

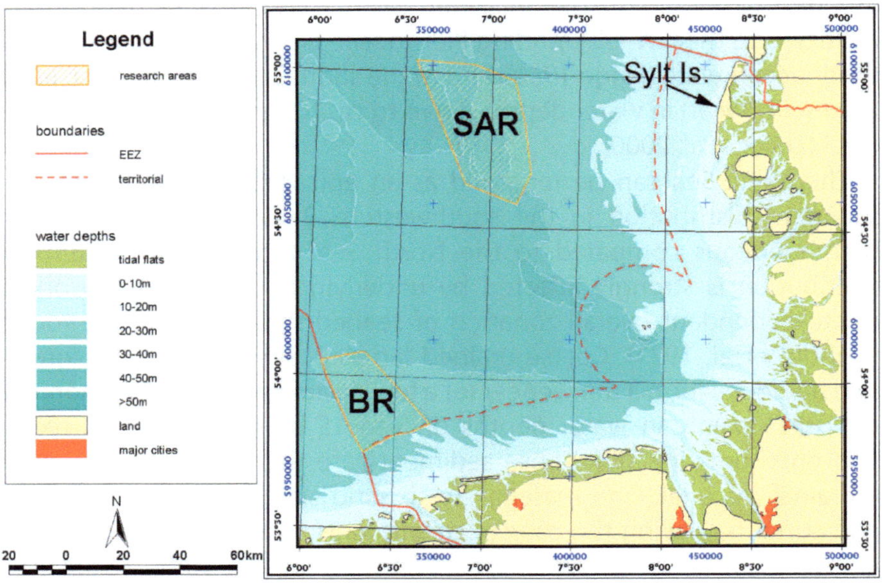

Figure 1. Location of the research area *Sylt Outer Reef* (SAR) in the German North Sea EEZ. BR denotes a further research area (*Borkum Reef Ground – Borkum-Riffgrund*) not discussed in this text

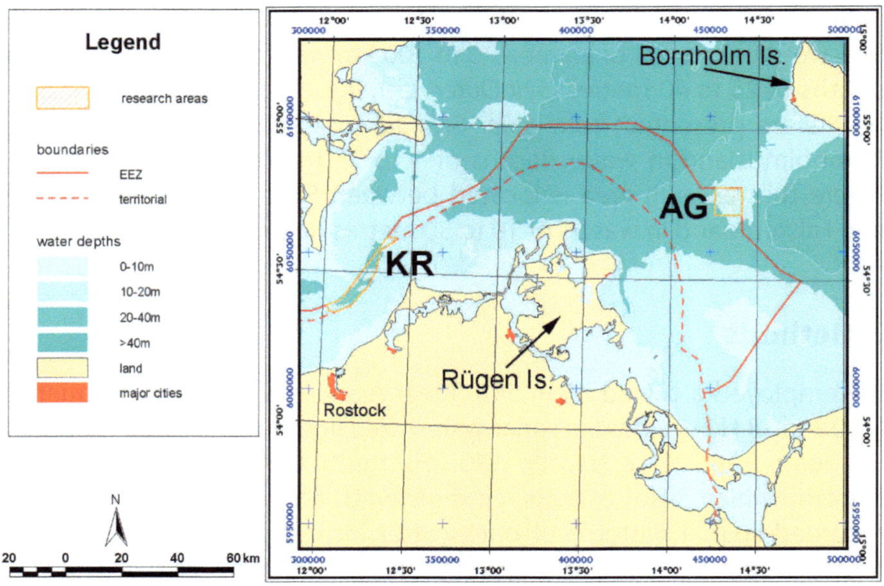

Figure 2. Location of the research areas *Kadet Trench* (KR) and *Adler Ground* (AG) in the German Baltic Sea EEZ

collected routinely. The ship's tracklines were chosen to allow full coverage of the seafloor. Positioning was achieved by differential GPS, which provided an accurate position of the vessel (and therefore also of the multibeam) within a 5-metre margin. The sidescan sonar was towed behind the vessel at a distance of about 20 to 50 metres, depending on the water depth and the range selected. This offset was accounted for by a constant value for each profile when calculating the position of the towfish[1]. The positioning error is thus greater compared to the multibeam, but this is difficult to quantify.

The Klein 595 sidescan sonar system was run in the high frequency mode in order to allow the highest resolution imaging. Due to the limited water depths, the selected ranges were 75 metres (Baltic Sea) and 100 metres (North Sea) on each side of the sidescan sonar. The 126 individual beams of the Seabeam 1185 multibeam system allowed a swath width[2] of about seven times the water depth.

Sidescan sonar data were recorded in digital format employing the Isis software package (Triton Elics International). These data were processed and geo-referenced[3] using the same software in order to create sidescan sonar mosaics of the study areas. Multibeam data were collected digitally with the program HydroStar (Elac Nautik). The post-processing of the raw data was conducted using the software HDPpost (Elac Nautik). The

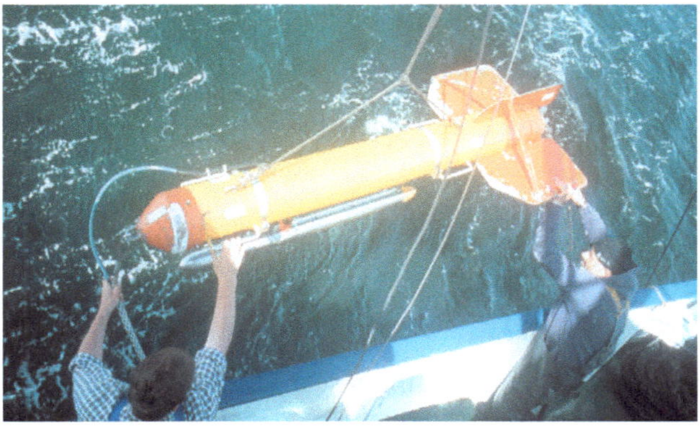

Figure 3. A sidescan sonar system was used (photo © Markus Diesing)

[1] Towfish: A streamlined body towed behind the vessel upon which sidescan sonar transducers are mounted and in which electronic modules are installed.
Transducer: The electromechanical component of a sonar system that is mounted underwater and converts electrical energy to sound energy and vice versa.

[2] Swath width: The lateral coverage of the sidescan sonar or multibeam echosounder on the seabed.

[3] Geo-referencing: Digitally attaching geographical position data to environmental sensor survey data.

mosaic files were displayed in the geographic information systems Delph Map (Triton Elics International) and Arc View (Esri). Backscatter strength is displayed as grey scale from light (low) to dark (high).

Single-beam bathymetry data were corrected for water-level changes by the Federal Maritime and Hydrographic Agency. In order to achieve a spatial distribution of water depth, the data were gridded including variogram[4] analysis.

Descriptive terms for grain size correspond to DIN 4022: clay: <2 µm; silt: 2–63 µm; fine sand: 63–200 µm; medium sand: 200–630 µm; coarse sand: 630 µm–2 mm; gravel: 2–63 mm; and boulders: >63 mm in diameter.

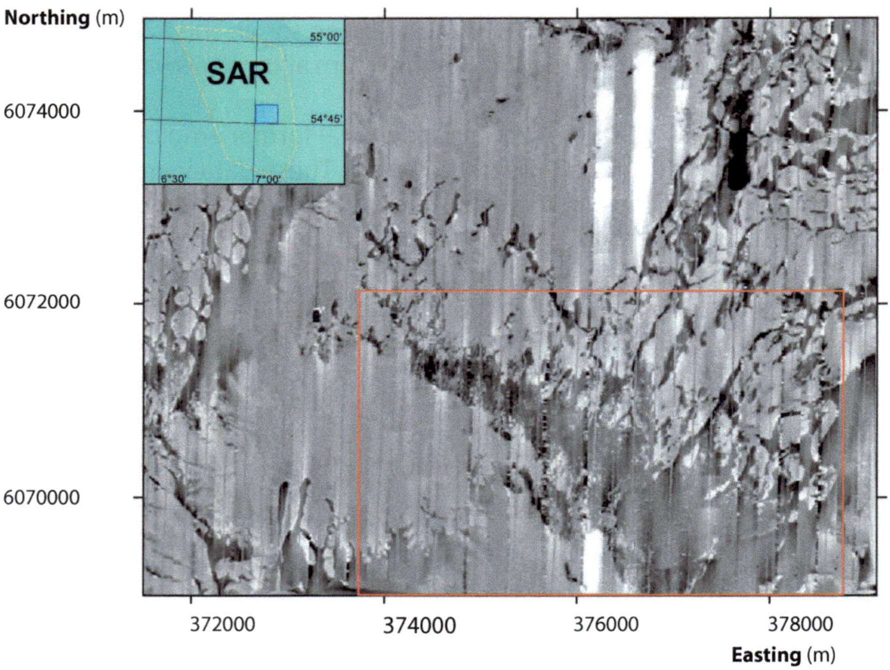

Figure 4. Multibeam backscatter image from the research area *Sylt Outer Reef*. The location of the displayed area is indicated by the blue rectangle in the inset (upper left corner). High backscatter intensity (dark grey) is indicative of coarse sands and gravels, low backscatter intensity (light grey) characterises areas covered with fine-to-medium sand. A strong spatial heterogeneity of seafloor sediments is clearly visible. White vertical stripes are artefacts. The red box indicates the location of figure 6.

4 Variogram: A measure of the variance between data as a function of distance.

4 Results and interpretation

4.1 German North Sea EEZ

We mapped a total of 315 km² of seafloor with the multibeam system within the research area *Sylt Outer Reef*. Figure 4 shows a typical example of the sediment distribution patterns and a clear image of the spatial heterogeneity of seafloor sediments. Based on different backscatter strength and ground-truthed by surface sediment samples and underwater video observations, three basic categories of seafloor sediments are distinguished: (a) coarse-grained sediment, (b) fine-to-medium sand, and (c) boulders.

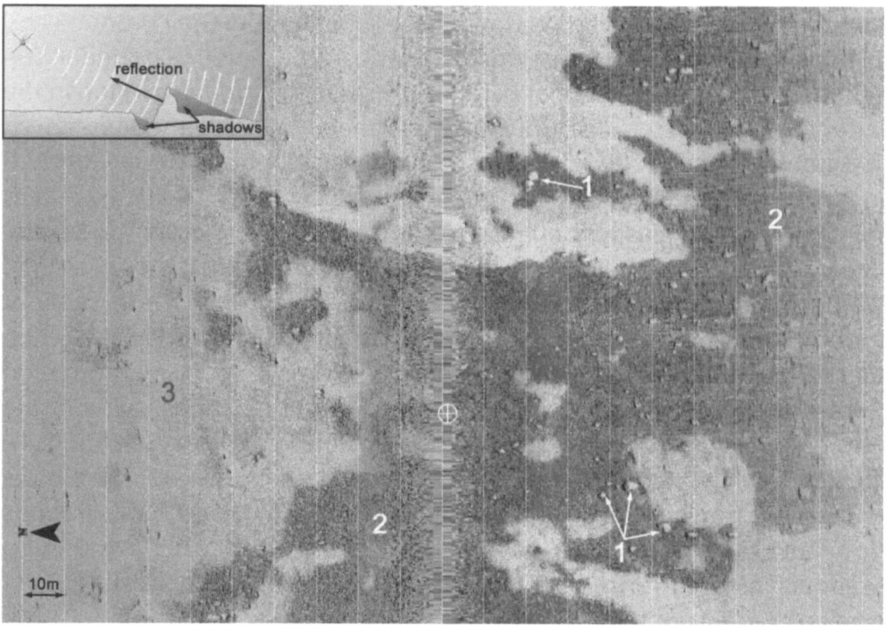

Figure 5. Geometrically corrected sonography of the seafloor from *Sylt Outer Reef*: Single boulders (1) can be identified on a patch of coarse sand (2). Several protruding boulders are visible within an area of fine-to-medium sands (3). The inset in the upper left corner explains how strong reflections and acoustic shadows behind obstacles (e.g., boulders) evolve; after Fish and Carr (1990), modified.

High and homogeneous backscatter (dark grey) is indicative of coarse sands and gravels. These sediments generally show ripple marks with wavelengths of up to ca. 1 metre. Co-registered multibeam bathymetry and backscatter data reveal that the patches composed of coarse material occur in depressions of several decimetres deep. Such sediments resemble "rippled scour depressions" (Cacchione et al. 1984) or "sorted bedforms" (Murray and Thieler 2004), which are widespread on sediment-starved continental shelves.

Low and homogeneous backscatter (light grey) areas are composed of fine-to-medium sands with low contents of mud and gravel (both below 5 weight-percent). Small-scale ripples in the order of a few centimetres wavelength were observed by underwater video. However, they are beyond the resolution of the employed sidescan sonar and multibeam systems.

Detailed investigations with high-resolution sidescan sonar additionally reveal the presence of boulders in some areas (figure 5). They are easily identified in the sonographies by the acoustic shadow they produce and which are interpreted as erosional lag deposits of morainal material. The boulders are found on or near patches of coarse-grained sediment. In the latter case, they protrude through a thin veneer of fine-to-medium sands. The boulders are not equally spaced on the seafloor but are concentrated in distinct areas. The number of individual boulders in such areas is relatively low. Typically, boulders are spaced a few meters (to several decametres) apart.

The relationship between different surface sediments and seafloor topography is visible on the sidescan sonar imagery, especially when merged with bathymetric data. In figure 6, a NW-SE-trending depression is visible. Maximum depth differences between the deepest parts of the depression and the highest parts of the surrounding seafloor are about 10 metres. Coarse-grained sediments are mainly found in the deepest parts and on the northeastern flank of the depression. Fine-to-medium sands dominate the southwestern flank of the depression and the surrounding seafloor. The occurrence of boulders is closely linked to the distribution of rippled coarse sediment.

4.2 German Baltic Sea EEZ

Within the *Kadet Trench* research area, the fieldwork was concentrated on the central part, where the Kadet channel cuts through the NW–SE-striking *Darss Sill* (*Darsser Schwelle*). In this area are found all types of surface sediments described above (see section 2) (figure 7). The largest part of the mapped area is covered by lag deposits, indicated by high and heterogeneous backscatter (dark grey). Their distribution is related to the general topography, that is, lag deposits are often found on the bathymetric highs. Lag deposits are comprised of sediment grain sizes from coarse sand to boulders, which may have diameters of more than 1 metre. Boulders are widely distributed within such lag deposits, but the density of boulders per unit seafloor area varies. Typically, the distance between individual boulders varies between a few decimetres and several meters. The density of boulders per unit seafloor area is therefore higher compared to the results gathered from the North Sea field site.

In the southwest, medium-to-low and homogeneous backscatter (light grey) indicates the presence of fine-to-medium sands. In the transition zones from sand to lag deposit, a strongly heterogeneous and patchy sediment distribution pattern is observed. A field of very large subaqueous dunes (according to the classification of Ashley 1990), with crest spacings in the range of 400 metres and crest heights of up to 5 metres, was detected on a shoal southwest of the investigated area (figure 7).

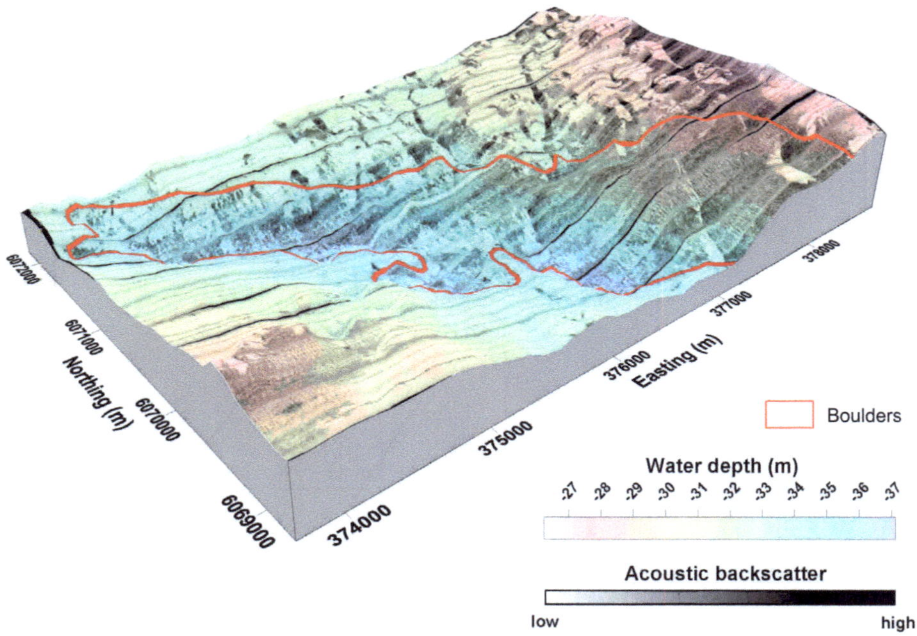

Figure 6. Three-dimensional image of the seafloor within the research area *Sylt Outer Reef*. View is from SW. Bathymetry is shown as 3D-relief and colour-coded water depths. Backscatter intensity is shown in greyscales, draped over the bathymetry. The occurrence of boulders is indicated by the red polygon. Vertical exaggeration: 75-fold.

A patch of low and homogeneous backscatter within the lag sediment area indicates the presence of organic-rich muddy fine sand to sandy mud. These sediments are deposited within a slight depression that favours the sedimentation of fine-grained material.

The research area *Adler Ground* is located at the southwestern tip of a major shoal (*Rønne Bank* (*Rönnebank*)). Here, the seafloor is widely covered by lag deposits, indicated by high and heterogeneous backscatter (figure

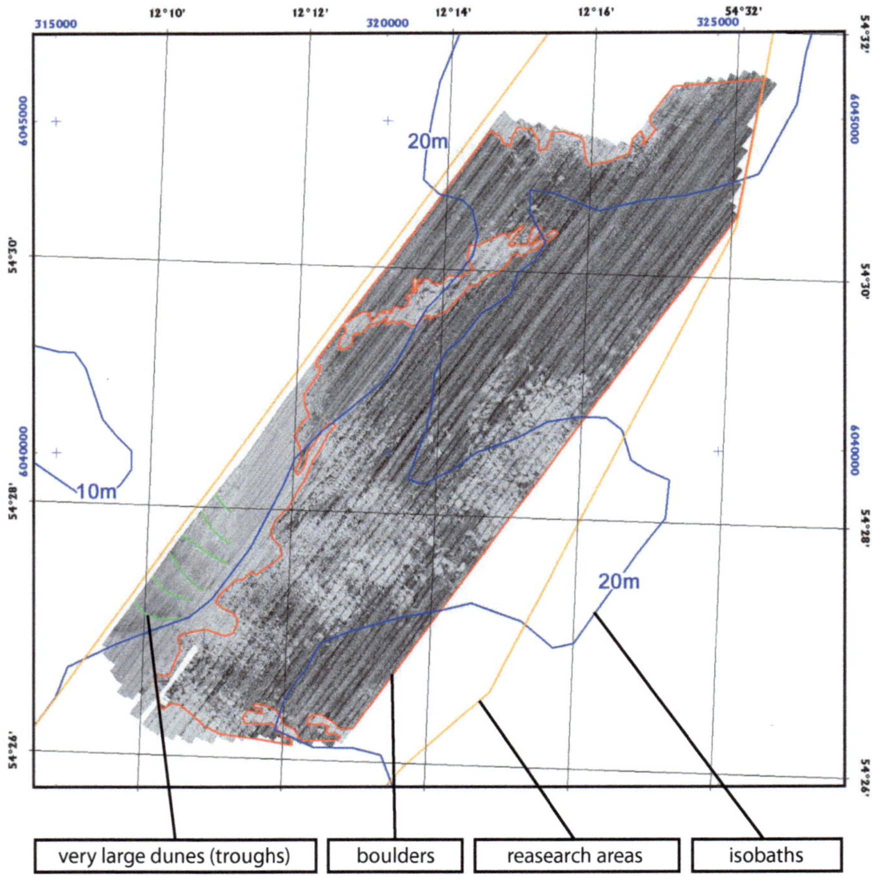

Figure 7. Sidescan sonar mosaic showing the centre part of the research area *Kadet Trench*. Lag deposits (dark grey) are widespread. The occurrence of boulders is closely linked to the distribution of lag deposits. Fine-to-medium sands (light grey) are restricted to the lower half of the image and are especially abundant in the SW, where they built up very large dunes

8). Seabed imagery as well as underwater-video data show small-scale changes within the lag deposits, alternating between boulders, gravel, and medium-to-coarse sands. Boulders are often populated by bivalves (*Mytilus sp.*). The density of boulders per unit seafloor is comparable to that in the *Kadet Trench* research area and is much higher than in the North Sea. There is no distinct relationship between the distribution of boulders and that of lag deposits. In some places, lag deposits are covered by a thin veneer of sand. In such situations, boulders may protrude through this cover of sand. Examples are found in the northeast of the investigated area (figure 8).

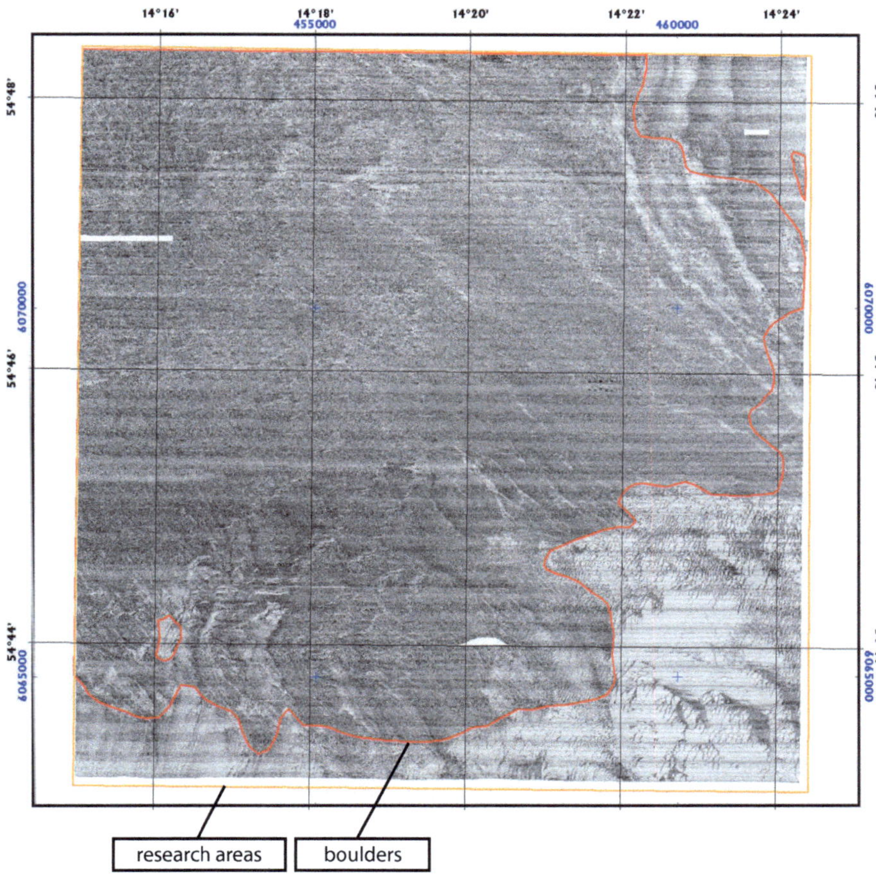

Figure 8. Sidescan sonar mosaic showing the research area *Adler Ground*. Lag deposits (dark grey) are widespread. The distribution of boulders is indicated by the red polygon. Fine-to-medium sands are restricted to the NE, SE, and SW of the image

Within the lag deposits, and in places where they are only covered by a thin veneer of sand, distinct morphological ridges, several metres wide and hundreds of metres long, were found (figure 9). Such ridges are densely covered by boulders, while the surrounding seafloor is often covered by rippled sand and gravel.

Fine-to-medium sands (light and homogeneous grey in the sonographies) dominate the northeast-, southeast-, and southwest corners of the research area. In the southeast, they alternate with NNE–SSW-directed strips of coarse sand. These features have spacings in the order of 50 metres (figure 8). Such strips of coarse sand resemble those at *Stoller Ground* (*Stoller Grund*), which have been interpreted as being transverse, ripple-like, and current-induced bedforms (Werner et al. 1976).

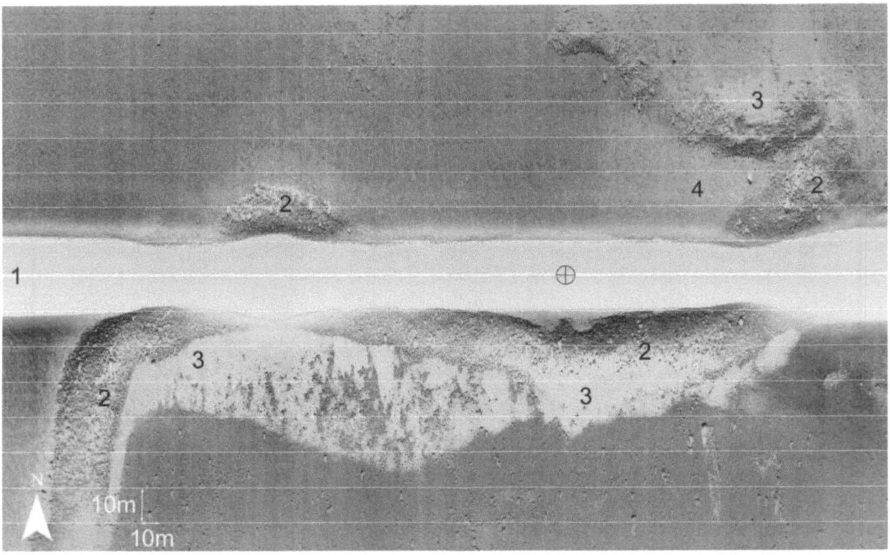

Figure 9. Uncorrected sonography of the seafloor from *Adler Ground*: 1 – water column, 2 – ridges, 3 – acoustic shadow, 4 – sand

5 Discussion

Boulders are present in both the German North Sea and the Baltic Sea as results of the modification of a glaciated terrain by marine transgression. Both areas were affected by glaciations. Consequently, morainal material was deposited. This sediment is comprised of grain sizes from clay to boulders of up to several metres in diameter. It is one of the most important source-sediments for modern redeposition and sedimentation. While the fine components were moved away due to marine reworking, the coarse fractions, especially boulders, remained almost where they were initially deposited. The resulting boulder fields are thus indications of former moraines.

From a practical point of view, this statement can also be read the other way around. If we are looking for submarine boulders, we should investigate continental shelf areas where moraines, especially terminal moraines, are present. For example, the *Darss Sill* is the submarine continuation of a Weichselian marginal (i.e., terminal moraine) line called Velgaster Staffel (Lemke et al. 1994, Lemke 1998). In fact, this area is characterised by the widespread occurrence of boulders (figure 7). In contrast, the situation is much less clear in the North Sea, where the knowledge is still too poor to reconstruct ice-marginal lines (Figge 1983). Here, the continental shelf was not affected by the last (Weichselian) glaciation. Glacial relicts are much older (>130,000 years ago), belonging to the Saalian and earlier glaciations. The time span available to rework, redistribute and level-out these glacial sediments by subaerial and subaqueous processes was thus much longer compared to that in the Baltic Sea, where the deglaciation started only ca. 15,000 years ago (Boulton et al. 2001). Moreover, waves, tides, and currents are much stronger in the North Sea and thus more effective as an erosional agent.

This strong contrast in the geological history of the two different continental shelf areas also explains why the density of boulders per unit seafloor area is much higher in the Baltic Sea, and why boulders in the Baltic are often linked to bathymetric highs while in the North Sea they are not. In the latter, we found boulders at one flank of slight depressions where relict coarse sediments are present (figure 6).

6 Conclusions

High-resolution acoustic seafloor-imaging techniques are suitable tools in order to map and characterise habitats of ecological importance such as boulders. Boulders are present in both formerly glaciated shelf areas, the German North Sea and the Baltic Sea. We observe differences regarding the distribution and density of boulders per unit seafloor between both areas. Such differences are explained by different geological evolution (i.e., timing of glaciation, subaerial exposure and marine transgression) and different intensity of hydrodynamic forcing (i.e., waves, tides, currents).

Acknowledgements

This work was funded by Bundesamt für Naturschutz and Bundesministerium für Umwelt, Naturschutz und Reaktorsicherheit (grant no. 802 85 270). Fieldwork was assisted by the captain and crew of the research vessel *Littorina* (Leibniz Institute for Marine Sciences, Kiel). Bathymetric data were corrected for waterlevel variations by the Federal Maritime and Hydrographic Agency, Hamburg. This paper benefits from the comments of three anonymous reviewers. We thank all persons and institutions mentioned above.

References

Ashley GM (1990) Classification of large-scale subaqueous bedforms: a new look at an old problem. Journal of Sedimentary Petrology 60:160–172

Boulton GS, Dongelmans P, Punkari M, Broadgate M (2001) Paleoglaciology of an ice sheet through a glacial cycle: the European ice sheet through the Weichselian. Quaternary Science Reviews 20 (4):591–625

Cacchione DA, Drake DE, Grant WD, Tate WB (1984) Rippled scour depressions on the inner continental shelf off central California. Journal of Sedimentary Petrology 54:1280–1291

Cochrane GR, Lafferty KD (2002) Use of acoustic classification of sidescan sonar data for mapping benthic habitat in the Northern Channel Islands, California. Continental Shelf Research 22 (5):683–690

Emery KO (1968) Relict sediments on continental shelves of world. American Association of Petroleum Geologists Bulletin 52 (3):445–464

Figge K (1980) Das Elbe-Urstromtal im Bereich der Deutschen Bucht (Nordsee). Eiszeitalter und Gegenwart 30:203–211

Figge K (1981) Sedimentverteilung in der Deutschen Bucht. Karte Nr. 2900, Deutsches Hydrographisches Institut, Hamburg

Figge K (1983). Morainic deposits in the German Bight area of the North Sea. In: Ehlers J (ed) Glacial deposits in North-West Europe. Balkema, Rotterdam, pp 299–303

Fish JP, Carr HA (1990) Sound underwater images. A guide to the generation and interpretation of side scan sonar data. Lower Cape Publishing, Orleans, MA

Fleming K, Johnston P, Zwartz D, Yokoyama Y, Lambeck K, Chappell J (1998) Refining the eustatic sea-level curve since the Last Glacial Maximum using far- and intermediate-field sites. Earth and Planetary Science Letters 163 (1–4):327–342

Kostylev VE, Todd BJ, Fader GBJ, Courtney RC, Cameron GDM, Pickrill RA (2001) Benthic habitat mapping on the Scotian Shelf based on multibeam bathymetry, surficial geology and sea floor photographs. Marine Ecology Progress Series 219:121–137

Lemke W (1998) Sedimentation und paläogeographische Entwicklung im westlichen Ostseeraum (Mecklenburger Bucht bis Arkonabecken) vom Ende der Weichselvereisung bis zur Litorinatransgression. Meereswissenschaftliche Berichte, 31

Lemke W, Kuijpers A, Hoffmann G, Milkert D, Atzler R (1994) The Darss Sill, hydrographic threshold in the southwestern Baltic: Late Quaternary geology and recent sediment dynamics. Continental Shelf Research 14 (7–8):847–870

Murray AB, Thieler ER (2004) A new hypothesis and exploratory model for the formation of large scale inner-shelf sediment sorting and "rippled scour depressions". Continental Shelf Research 24 (3):295–315

Niedermeier-Lange R, Werner F (1988) Flachseismische und sonographische Aufnahmen im Küstenvorfeld der Hohwachter Bucht (westliche Ostsee) und ihre glazialgeologische Interpretation. Senckenbergiana maritima 20:59–79

Ojeda GY, Gayes PT, Van Dolah RF, Schwab WC (2004) Spatially quantitative seafloor habitat mapping: example from the northern South Carolina inner continental shelf. Estuarine, Coastal and Shelf Science 59 (3):399–416

Schulz HD, Tauchgruppe Kiel (1983) Der Steingrund bei Helgoland – Restsediment einer saalezeitlichen Endmoräne. Meyniana 35:43–53

Seibold E, Exon N, Hartmann M, Kögler F-C, Krumm H, Lutze GF, Newton RS, Werner F (1971) Marine geology of Kiel Bay. In: Müller G (ed) Sedimentology of parts of Central Europe. Guidebook. VIII. Int. Sediment. Congress, Frankfurt a.M. 1971, pp 209–235

Tahrir S (1984) Oberflächensedimente und geologischer Bau des flachen Untergrundes im Gebiet des Kriegersflach (Mittlere Ostsee). Reports, Geol.-Paläont Inst Univ Kiel, 8

Tauber F, Lemke W (1995) Map of sediment distribution in the western Baltic Sea (1:100,000), Sheet "Darß". – Deutsche Hydrographische Zeitschrift 47 (3):171–178

Tauber F, Lemke W, Endler R (1999) Map of sediment distribution in the western Baltic Sea (1:100,000), Sheet Falster – Møn. Deutsche Hydrographische Zeitschrift 51 (1):5–32

Werner F (2004) Coarse-sand patterns in the south-eastern German Bight and their hydrodynamic relationships. Meyniana 56:117–136

Werner F, Arntz WE, Tauchgruppe Kiel (1974) Sedimentologie und Ökologie eines ruhenden Riesenrippelfeldes. Meyniana 26:39–62

Werner F, Altenkirch J, Newton RS, Seibold E (1976) Sediment patterns and their temporal variation on abrasion ridges in a moderate flow regime. Meyniana 28:95–105

Winn K, Werner F (1984) Ein integriertes akustisches Fernmeßsystem (IAFMS) zur Typisierung von Schichtfolgen im Meeresboden und Bestimmung ihrer Raumlage. - Berichte - Reports, Geol.-Paläont Inst Univ. Kiel, 5

Winn K, Averdieck F-R, Werner F (1982) Spät- und postglaziale Entwicklung des Vejsnaes-Gebietes (Westliche Ostsee). Meyniana 34:1–28

Zeiler M, Schulz-Ohlberg J, Figge K (2000) Mobile sand deposits and shoreface sediment dynamics in the inner German Bight (North Sea). Marine Geology 170 (3–4):363–380

Chapter 7

Search for particularly valuable benthic areas within the German North Sea EEZ*

Eike Rachor

Alfred-Wegener-Institute for Polar and Marine Research, Bremerhaven

Abstract

Following an extensive bottom animal (zoobenthos) sampling expedition in early summer 2000, nine different faunal assemblages (communities) have been identified in the sublittoral offshore waters from the North of the *Doggerbank* to the inner parts of the *German Bight* (*Deutsche Bucht*). The characteristics of these communities are presented, and their specific values for protection measures are indicated. Special attention is given to complex biotope areas where rare species and rich mosaics of communities occur. Since many of these areas fulfil the ecological criteria for *sandbank* and *stone reef* habitats of the European Habitats Directive, they have been proposed for site protection (see figure 1 in chapter 4).

1 Introduction

Macrozoobenthos[1] is a good indicator of seafloor environmental conditions. Its infaunal components[2] can be studied quantitatively, thus allowing a statistically derived delineation of bottom communities. Since the beginning of the 20th century, there have been various but scattered studies on North Sea benthos,[3] which became quantitative after the work of Petersen (1911). In 1986, a large-scale survey of the North Sea by members of the International Council for the Exploration of the Sea (ICES) Benthos Ecology Working Group produced an overall description of the

---

* This is AWI-publication n 15393.

1 Macrozoobenthos: larger seafloor animals.

2 Infaunal components: living within the sediment.

3 Benthos: seafloor organisms.

North Sea sublittoral[4] zoobenthos distribution (Künitzer et al.1992, Heip et al. 1992). It showed general depth-related and latitudinal trends in main parameters like biomass and diversity, and sediment relationships within this general trend scheme. The ideas of Glémarec (1973) of community zonation according to seafloor depth étages,[5] which indicate the strong influences of temperature and mixing conditions, were substantiated by the survey results.

A few years earlier, Salzwedel et al. (1985) presented a detailed description of seafloor communities in the *German Bight*, which allowed comparisons with earlier studies, especially those of Hagmeier (1925), Stripp (1969), and Dörjes (e.g., 1977). These descriptions show clear relationships between distribution patterns of faunal assemblages and sediment conditions, water depths and distances from the shore. Rachor (1990) explained changes in the zoobenthos of the *German Bight* between the 1920s and 1970s (and later) mainly through the influences of increased eutrophication.[6] Today, bottom fisheries are regarded as main additional impacts on the North Sea benthos. Further, some shifts in the distribution patterns of species seem to be related to climatic trends (e.g., warming). Other studies have contributed to the understanding of zoobenthos distribution and its temporal changes especially in the *Doggerbank* area (see e.g., Kröncke 1992).

In 2000, a new survey was performed in and around the German Exclusive Economic Zone (EEZ) of the North Sea by the author's research group in close cooperation with the group of Kröncke (Wilhelmshaven). It was at the same time a contribution to a new North Sea-wide benthos survey within the ICES framework (Rees et al. 2002). This work was initiated and strongly supported by the German Federal Agency for Nature Protection (BfN). It aimed to search for and identify important biotopes and communities in need of protection, especially as contribution to the European NATURA 2000 network of protected areas (Habitats Directive 1992).

This article presents a summary of the work in 2000 and later. The focus is on the German EEZ (see Rachor and Nehmer 2003).

---

[4] Sublittoral: zone between the low-tide mark and the edge of the continental shelf (here about 200m depth).

[5] Étages according to Glémarec are general marine zonations of bottom-inhabiting biocoenoses depending on the temperature regime in different water depths (as a rule, depths zonations).

[6] Eutrophication: increased availability of plant nutrients, such as nitrates and phosphates, which stimulate growth of algae and of organisms feeding on them, sometimes followed by oxygen depletion in the water.

2 Area of investigation and methods

The German North Sea EEZ covers the southwestern North Sea, including the *German Bight* and part of the *Doggerbank*, up to the central North Sea. These mainly sandy-to-muddy areas with depths between 10 and 70 metres were sampled from 30 May to 24 June 2000. Some neighbouring parts of the North Sea, especially those in Dutch, Danish, and UK waters, were also sampled at the same time.

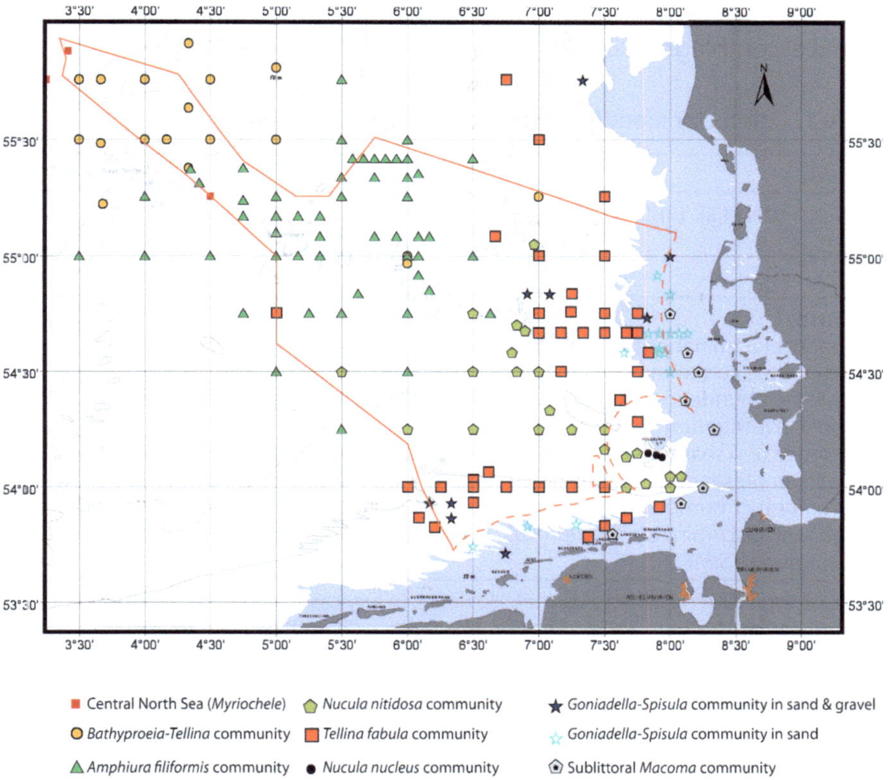

Figure 1. Distribution map of main sublittoral macrozoobenthos communities in the EEZ of the German North Sea and its surroundings in 2000. Light blue stars: medium-to-coarse sand variant of the *Goniadella-Spisula* community (c-m in table 2, 5th column). Dark blue and black: coarse-sand/gravel variant (c-g in table 2) of the *Goniadella-Spisula* community. This relict sediment variant is often associated with stony grounds/reefs and their typical epifauna. The sublittoral *Macoma balthica* community and neighbouring parts of the *Tellina fabula* community are displayed only in parts. Some information about distributions (mainly of the relict sediment *Goniadella-Spisula* and stone reef epifauna communities) is added according to own findings in other years, especially in 2001–2003 (black stars).

Infauna[7] was collected mainly by a 0.1m² van Veen grab (2 grabs per station), sieved on board over a 1.0-mm screen, and fixed and preserved in borax-neutralised 5% formaldehyde. Sorting, countings, weightings (wet total biomass) and identifications were done in the laboratory. Epifauna[8] was obtained by a 2-metre beam trawl and a larger otter trawl. Additionally, underwater videos and photographs were taken by towed camera systems and divers.

Following laboratory analyses, cluster- and multidimensional scaling (MDS) procedures were applied to the infauna data, using the PRIMER 5 software package (Plymouth, UK 2000), to identify and delineate benthic faunal assemblages (communities) and their biotopes.

Table 1. Criteria for identifying characteristic bottom fauna species

Criteria	ND	FA	P	FP	RDS
In Salzwedel et al. 1985	> 5%	> 66%	> 80%	> 66%	–
This paper	not < 3%	> 60%	> 70%	> 60%; not < 40%	ranks 1–5

Criteria

- numerical dominance (ND = abundance of a species / total abundance of all species in the same community);
- fidelity in abundance (FA = total abundance in a community / total abundance in the whole investigation area);
- presence (P = share of stations in a community where the species was found);
- fidelity in presence (FP = number of stations where the species was found within a community / number of stations with its findings in the whole investigation area);
- high rank discriminator species (RDS, responsible for dissimilarities of clusters according to SIMPER).

At least three criteria have to be met. However, in any case, numerical dominance (ND) should not be less than 3%, and fidelity in presence (FP) not less than 40%. Based on terrestrial methods, earlier authors used such combinations of criteria, too, but put more emphasis on biomass dominance and presence (e.g., Thorson 1957; see also Stripp 1969).

Characteristic species of bottom communities were identified according to the modified and extended procedure described by Salzwedel et al. (1985), who used numerical dominance, fidelity in abundance, presence and its fidelity as selection criteria. Species important for the separation of main clusters (assemblages) according to SIMPER analyses (PRIMER 5 software) were additionally included. Criteria for the selection of characteristic species are presented in table 1.

[7] Infauna: animals living in bottom sediments.

[8] Epifauna: animals living on the seafloor or on specific substrate surfaces, e.g., on stones.

The epifauna data were used as additional qualitative information in describing communities, especially for delineations of stony, reef-like biotopes and their fauna. For more details, consult the original research report (Rachor and Nehmer 2003).

3 Results

Nine different sublittoral zoobenthos assemblages were identified in the offshore area from the inner *German Bight* up to the central North Sea, north of the *Doggerbank* (table 2 and figure 1). Their distributions correspond with the main morphological and sedimentological conditions in the area (i.e., water depths, distance from the coast, sediment grain size and heterogeneity of the substrate).

The *Nucula nucleus* community was only represented in a few single samples from the *Helgoland Deep Trench* (*Helgoländer tiefe Rinne*). The next rarest assemblages (i.e., stony reef epifauna and coarse sand (or *Goniadella-Spisula*) infauna communities were normally restricted to relict sediments (such as moraine remainders of boulders, stones, gravel and coarse sands). Such biotopes were found scattered in the *Eastern German Bight* (*Östliche Deutsche Bucht*) (especially on and near the eastern slope of the Pleistocene Elbe valley), on and around the *Borkum Reef Ground* (*Borkum-Riffgrund, Western German Bight* (*Westliche Deutsche Bucht*)), and, to a minor extent, also in certain parts of the *Doggerbank*.

With the exception of the *Nucula nitidosa* community, which expanded to the northwest (into the former area of the *Amphiura filiformis* community), the distributions of the infauna assemblages were generally the same as already described by Salzwedel et al. (1985).

Typical and characteristic species are presented in table 2.

Notes to accompany Table 2 overleaf

Explanations: Typical species comprise species which are characteristic (*) as well as species of high dominance and fidelity.

SN = average species number (2 grabs, together 0.2 m^2); RL = total number of Red List species; N = total number of grab stations; c-m = coarse-to-medium sand variant, c-g = coarse-sand-to-gravel variant of the *Goniadella-Spisula* community

Comments: *Nucula nitidosa* community: In 2000, it also comprised certain parts of the Pleistocene Elbe valley which was classed as *Amphiura* community in earlier years. *Tellina fabula* community: the clam *Chamelea gallina* (syn. *Venus gallina, V. striatula*) is no longer regarded characteristic for fine-sand communities (as in Stripp 1969), but is nowadays rather typical for the *Amphiura* community.

Table 2. Sublittoral zoobenthos communities in the EEZ of the German North Sea (central to southeastern North Sea)

No. and abbreviation	Zoobenthos communities	Synonyms	Occurrences
1 Tel fab	***Tellina fabula***	*Tellina fabula* community (Salzwedel et al. 1985); *Venus gallina* association (Stripp 1969; Hagmeier 1925)	*Eastern* and *Western German Bight* (G.B.) N = 39
2 Gon Spi	***Goniadella-Spisula***	*Goniadella-Spisula* comm. (Salzwedel et al. 1985); "impoverished" variant of the *Venus gallina* ass. (Stripp 1969); Coarse sand comm. (Dörjes 1977)	*Eastern* and *Western G.B.* and near Helgoland N = 25
3 Bat Tel	***Bathyporeia-Tellina***	*Tellina fabula* comm. (Kröncke u.a. 1991); *Venus gallina* comm. (Ursin 1952; Birkett 1953)	*Doggerbank* N = 21
4 Amp fil	***Amphiura filiformis***	*Echinocardium cordatum-Amphiura filiformis* ass. (Stripp 1969; Hagmeier 1925)	*Central G.B.* (from the Pleistocene Elbe valley to the NW) N = 55
5 Mac bal	***Macoma balthica***	*Macoma balthica* ass. (Stripp 1969, and others)	Near to inshore waters N = 8
6 Myr cNS	***Myriochele* (central NS)**		Central North Sea N = 6
7 Nuc nit	***Nucula nitidosa***	*Abra (Scrobicularia-)alba* ass. (Stripp 1969; Hagmeier 1925)	In front of Elbe and Weser Estuaries N = 24
8 Nuc nuc	***Nucula nucleus***	*Nucula nucleus* comm. (Caspers 1938)	Trench south of *Helgoland* (*"Helgoländer tiefe Rinne"*) N = 3
9 St-Fauna	**Fauna of Stones and Stone Reefs**	(Fauna of stony grounds, Kühne and Rachor 1996)	*Eastern* and *Western G.B.* and near Helgoland (N = few)

Characteristics	Typical species (characteristic: *)
Fauna of fine-to-medium sands in the sublittoral off the Wadden Sea at 15 to 30 m depths; SN = 27 RL = 25	*Tellina fabula**, *Magelona johnstoni**, *Urothoe poseidonis**, *Bathyporeia guilliamsoniana**, *Spiophanes bombyx, Goniada maculata*
Fauna on moraine banks (coarse-sands-to-gravel). Two variants are distinguished: on coarse-sands-to-gravel and on coarse-to-medium sands; SN = 14 RL = 22	*Aonides paucibranchiata** (c-g), *Ophelia limacina** (c-m), *Thracia villosiuscula** (c-m), *Polygordius* spp. (c-g), *Echinocyamus pusillus* (c-g), *Branchiostoma lanceolatum** (c-g), *Spisula* spp.* (c-m), *Goniadella bobretzkii* (c-m), *Goodallia triangularis* (c-m)
Fine-sand fauna of the *Doggerbank*; SN = 44 RL = 29	*Bathyporeia** spp.: *B. elegans, B. nana, Amphiura brachiata**, *Spiophanes bombyx, Tellina fabula*
Fauna of muddy substrates, with some admixtures of fine-to-very-fine sands; missing in front of the estuaries; SN = 32 RL = 32	*Amphiura filiformis**, *Mysella bidentata**, *Harpinia antennata**, *Corbula gibba, Pholoe baltica*
Sublittoral variant adjacent to the Wadden Sea; SN > 10 RL > 9	*Urothoe poseidonis**, *Ensis americanus**, *Macoma balthica, Lanice conchilega*, partly: *Abra alba*
Fauna of muddy mixed substrates at greater depths of the central North Sea; SN = 35 RL = 18	*Myriochele* spp. *, *Paramphinome jeffreysii**, *Cerianthus lloydii, Abra prismatica*
Fauna of muddy substrates mainly in front of the estuaries; SN = 28 RL = 18	*Nucula nitidosa**, *Abra alba**, *A. nitida, Scalibregma inflatum**, *Phaxas pellucidus, Amphiura brachiata, Ophiura albida*
Fauna of mixed substrates (coarse sands with mud and shells); SN = 35 RL = 22	*Nucula nucleus* (former *), *Timoclea ovata, Nephtys kersivalensis, Amphipholis squamata, Scalibregma inflatum*
Epifauna, rich in species, living on stony grounds and reefs (relicts of [terminal] moraines); as a rule, closely allied with Gon Spi and Tel fab, forming rich mosaic community complexes RL > 20	*Metridium senile, Alcyonium digitatum, Flustra* spp., *Ophiothrix fragilis, Echinus esculentus, Cancer pagurus, Ascidiacea*

Table 3. Evaluation of the zoobenthos communities in the EEZ of the German North Sea from a nature conservation view

No. and abbreviation	Zoobenthos communities	Diversity (species numbers)		(RL species)	Natural-ness	Dependency of the species	
		total	mean	total			"excl."
1 Tel fab	***Tellina fabula***	131	27.1	25	+	+	7 (0.02)
2 Gon Spi	***Goniadella-Spisula***	101	13.7	22	++	++	14 (0.13)
3 Bat Tel	***Bathyporeia-Tellina***	146	43.9	29	+	++	15 (0.04)
4 Amp fil	***Amphiura filiformis***	173	31.5	32	+	+ to ++	42 (0.12)
5 Mac bal	***Macoma balthica***		> 10	> 9	+	+	4 (0.10)
6 Myr cNS	***Myriochele* (central NS)**	93	35	18	+	+ to ++	16 (0.17)
7 Nuc nit	***Nucula nitidosa***	104	27.9	18	–	+	4 (0)
8 Nuc nuc	***Nucula nucleus***	> 132	35	22	+	++	17 (0.6)
9 St-Fauna	**Fauna of Stones and Stone Reefs**	30 (epi)		(> 12) epi	+++	+++	> 30

Notes

+ = given; ++ = high; +++ = very high; - = disturbed (e.g., by demersal fisheries and/or eutrophication);

* general pollution, general eutrophication and fisheries' influences not indicated; RL = Red Lists; HD = from European Habitats Directive; "excl." = species exclusively found in this community (same column, number in brackets = "exclusive" Red List species per station)

Diversity here restricted to species number and to numbers of red list (RL) species (see Rachor et al. 1998). **Naturalness** lack of disturbances or degradation. **Dependency** the degree to which species and functions depend on an area/biotope ("key areas") and are exclusively/mainly found there, at least for some time of the year.[9] **Uniqueness, rareness** restriction of biotopes and

Uniqueness	Integrity of production, food web	Sensitivity, vulnerability	HD biotopes ("habitats")	Significance for German EEZ	Need of protection (P), specific threats (T)
+	+ to -	+	sandbank	+	P: part areas
+++	+	+++	sandbank and reef	++	P: high T: sand & gravel exploitation
++	+ to -	++	sandbank	++	P: transnational need
+	+ to -	+	–	+	P: part areas
+	+ to -	+	sandbank	++	P: sandbanks (coastal waters)
++	+ to -	+	–	++	P: in connection with 3
++	-	++	–	++	P: part area (coastal waters) T: eutrophication
+++	+ to -	++	–	++	P: high (coastal waters)
+++	+	+++	reef or reef-like	+++	P: high T: gravel exploitation

communities to only one or very few, or/and small areas. **Integrity of production, food web** undisturbed, functioning in a way that the community appears to be a somehow self-sustaining entity. **Sensitivity, vulnerability** susceptibility to degradation by natural events (disturbances) and/or human activities. **Significance for German EEZ** the degree to which a community/biotope represents natural (undisturbed) characters, is rare and important for protection

[9] Compare with definition of *aggregation* in Derous et al. (in preparation): "Degree to which an area is a site where major shares of individuals of a species are aggregated for some parts of the year, or a site which most individuals use for some important function in their life history, or a site where some structural property or ecological process occurs with exceptionally high density."

4 Evaluation and discussion

Table 3 is a summary view on the invertebrate zoobenthos communities around the German North Sea EEZ indicating their values, some threats to, as well as needs for nature conservation. Most of the evaluation criteria are explained in Rachor and Günther (2001), and to a large extent are derived from Salm and Price (1995). An extensive discussion of biological evaluation criteria will be published by S. Derous et al. (in preparation), where the main condensed criteria are *rarity*, *aggregation* (including dependency), *fitness consequences* (partly including integrity of functions), *biodiversity* and, possibly, *resilience/vulnerability*, while *naturalness* and *proportional importance* (*significance* for an area) are proposed as modifying criteria.

Scoring as done in table 3 is always somehow subjective. However, for *naturalness*, for example, the highest score (+++) was given to stony reefs, where demersal fishing is regarded as having no direct influence and where *dependency* is highest in cases in which species are totally dependent on the habitat. *Rareness* considers the frequency of occurrence and the spatial extension of a biotope in the investigated area, and *food web integrity* was considered disturbed in most cases mainly by eutrophication and fisheries. *Vulnerability*, although strongly correlated to *rarity*, considers also the location of the biotope, e.g., proximity to the main shipping routes, distance to potentially disturbing human activities, regeneration potential together with distance to similar biotopes and their "stepping-stone" function during dispersal/migration.

Another aspect to be considered is the potential role of (protected) biotopes (and populations) for the regeneration of disturbed or destructed biotopes and communities in other areas. During larval dispersal and migration such biotopes may serve as refuges and stepping-stones which safeguard coherence of species habitats. In order to provide such coherence functions, (protected) biotopes (and populations) should be less than ca. 100 km apart in the southeastern North Sea (see Rachor and Günther 2001).

While the widespread communities (of *Tellina fabula*, *Amphiura filiformis*, *Macoma balthica*, *Bathyporeia-Tellina* and *Nucula nitidosa*) as such need not be primarily considered for protection, rare and/or geographically restricted assemblages, such as *Nucula nucleus*, *Myriochele*, *Goniadella-Spisula* and the epifauna assemblage of stones and boulders (stony reefs) need special attention at least from a regional (German) perspective. The latter two communities (i.e., *Goniadella-Spisula* and epifauna assemblages) also need special attention from a broader (i.e., European) perspective since they are representative of the offshore biotopes (habitats) of sandbanks and reefs. From the European point of view, the *Bathyporeia-Tellina* community of the *Doggerbank* also has to be considered for protection, because this particular area comprises shallow as well as deeper parts of an offshore sandbank. Moreover, the above-mentioned *Goniadella-Spisula*, the epifauna as well as the *Bathyporeia-Tellina* communities (i.e., those of terminal moraine remainders and partly that of the *Doggerbank*) are normally closely combined with others in biotope complexes (i.e., mosaics) forming community mixtures, which then increase the species diversity in the areas. Such areas with relict sediments of gravel and stones are derived from remainders of (terminal) moraines and were found mainly on the *Borkum Reef Ground* in the west of the *German Bight*, on the eastern slope of the Pleistocene Elbe valley (*Sylt Outer Reef* (*Sylter Außenriff*)), and from the northwest (*Störtebecker Ground* (*Störtebecker Grund*)) to the east (*Stone Ground* (*Steingrund*)) of the island Helgoland. These biotope complexes also occur in smaller grounds of the *Eastern German Bight* (e.g., *Amrum Outer Ground* (*Amrumer Außengrund*)) and on the investigated part of the *Doggerbank*. Main parts of these complex biotope areas have been proposed for protection (see figure 1 in chapter 4).

Stone and especially boulder fields are obstacles for heavy bottom trawling gear of fishermen. Therefore their biotopes host more rare and long-lived species than muddy and sandy biotopes which are subject to very intensive bottom beam and otter trawling. In areas where heavy bottom fishing activities take place regularly, several epifauna and many long-lived species cannot build up populations with a considerable proportion of adults. Especially heavy beam trawls of boats fishing for sole can harm even the deep-dwelling infauna. Moreover, wide areas of fine-to-medium sands at depths of less than 30 metres can even be strongly affected during heavy storms. The sediment mobility of these areas may prevent a "climax-like" community (without succession and with a clearly defined species dominance) to develop (Rachor and Gerlach 1978). Fine-to-medium sand patches between stone fields are however, somehow protected against erosion and can thus sometimes be richer in specific organisms than in extended soft bottom areas.

In table 4 all these "advantages" of biotope complexes can be seen, if, for example, the average species numbers (total species number of such areas) are compared with the average numbers of the widespread larger communities (*) given in table 2.

Table 4. Macrozoobenthos, invertebrate species richness data (from 2000 and later, Rachor and Gusky 2004)

	Total species number		Average species number
	2000	2000–03	2000
Complex Biotope Area			
Borkum Reef Ground	122	276	9.4
Amrum Outer Ground	120	–	7.1
Sylt Outer Reef	119	285	10.8
Doggerbank	219	–	11.5
Community			
Tellina fabula community *	131	–	3.4
Goniadella-Spisula community	101	–	7.2
Bathyporeia-Tellina community	146	–	8.1
Amphiura filiformis community *	173	–	3.2
Nucula nitidosa community *	104	–	4.7

Figure 2. The hydroid *Tubularia* sp. at *Sylt Outer Reef*

5 Conclusions

In the German North Sea EEZ, the biotope complexes given in table 4 have been proposed for inclusion in a European network of marine protected areas from a benthic point of view. This is because these areas comprise the rarest offshore communities in the German EEZ and – through the combination of several assemblages – they allow the conservation of the highest biological diversity in comparatively small areas. All these complex areas at least partly fulfil the ecological criteria of subtidal sandbanks and stone reefs (European Commission 1999). They have to be protected according to the European Habitats Directive. In addition, their distribution in space enables them to function as stepping-stones and refuges for migrating, fluctuating, and endangered species. A great part of the complex biotope areas given in this paper also overlap with the sites identified by the German Federal Agency for Nature Protection as important especially for the harbour porpoise (chapter 11) and seals (chapter 10), as well as for birds (chapter 13).

Acknowledgements

Petra Nehmer, from Oldenburg, did the main species identification and community description work which I appreciate very much. I am also thankful to Elke Barwich and Manuela Gusky, from Bremerhaven, who supported the identification and editing work.

This study was initiated and strongly supported by the German Federal Agency for Nature Protection (BfN).

References

Birkett L (1953) Changes in the composition of the bottom fauna of the Dogger Bank. Nature 171:265

Caspers H (1938) Die Bodenfauna der Helgoländer Tiefen Rinne. Helgoländer wiss Meeresunters 2:1–112

Derous S, Agardy T, Hillewaert H, Hostens K, Jamieson G, Lieberknecht L, Mees J, Moulert I, Olenin S, Paelinckx D, Rabaut M, Rachor E, Roff J, Stienen E, van der Wal JT, Van Lancker V, Verfaillie E, Vincx M, Weslawski JM, Degraer S (in preparation) Biological valuation: Towards a scientifically acceptable and generally applicable strategy for the marine environment. Submitted to Aquatic Conservation.

Dörjes J (1977) Über die Bodenfauna des Borkum Riffgrundes (Nordsee). Senckenbergiana marit 9:1–17

European Commission (1999) Interpretation manual of European Union habitats. Version Eur 15/2. European Commision, DG Environment, Brüssel

European Council Directive 92/43/EEC (1992) 'Habitats/FFH Directive': On the conservation of natural habitats and of wild fauna and flora. Official Journal L 206 (22.07.1992), pp 07-050; amended by 97/62/EEC, Off. J. L 305, 1997, pp 42–65 (re. Annexes I–II)

Glémarec M (1973) The benthic communities of the European North Atlantic continental shelf. Oceanogr Mar Biol Ann Rev 11:263–289

Hagmeier A (1925) Vorläufiger Bericht über die vorbereitenden Untersuchungen der Bodenfauna der Deutschen Bucht mit dem Petersen-Bodengreifer. Ber Dt Wiss Komm Meeresforsch Neue Folge Bd 1:247–272

Heip C, Basford D, Craeymeersch JA, Dewarumez JM, Dörjes J, Wilde P de, Duineveld G,

Eleftheriou A, Herman PMJ, Niermann U, Kingston P, Künitzer A, Rachor E, Rumohr H, Soetaert K, Soltwedel T (1992) Trends in biomass, density and diversity of North Sea macrofauna. ICES J Mar Sci 49:13–22

Kröncke I (1991) The macrofauna distribution on the Dogger Bank in April/May 1985-87. Berichte Biol Anst Helgoland 8:1–137

Kröncke I (1992) Macrofauna standing stock of the Dogger Bank. A comparison: III. 1950–54 versus 1985–87. Helgoländer Meeresunters 46:137–169

Kühne S, Rachor E (1996) The macrofauna of a stony sand area in the German Bight (North Sea). Helgoländer Meeresunters 50:433–452

Künitzer A, Basford D, Craeymeersch JA, Dewarumez JM, Dörjes J, Duineveld GCA, Eleftheriou A, Heip C, Herman P, Kingston P, Niermann U, Rachor E, Rumohr H, de Wilde PAWJ (1992) The benthic infauna of the North Sea: species distribution and assemblages. ICES J Mar Sci 49:127–143

Petersen CGJ (1911) Valuation of the sea I. Animal life of the sea bottom, its food and quantity. Rep. Danish Biol Stat 20:1–81

Rachor E (1990) Changes in sublittoral zoobenthos of the German Bight with regard to eutrophication. Neth J Sea Res 25:209–214

Rachor E (1998) Rote Liste der bodenlebenden wirbellosen Meerestiere. Schr R Landschaftspfl Natursch 55:290–300

Rachor E, Gerlach SA (1978) Changes of macrobenthos in a sublittoral sand area of the German Bight, 1967 to 1975. Rapp P-v Réun Cons Int Explor Mer 172:418–431

Rachor E, Günther C-P (2001) Concepts for offshore nature reserves in the southeastern North Sea. Senckenbergiana maritima 31:353–361

Rachor E, Gusky M (2004) Benthologische Arbeiten zur ökologischen Bewertung von Windenergieanlagen-Eignungsgebieten in der Nordsee. Final report for BfN, Bremerhaven

Rachor E, Nehmer P (2003) Erfassung und Bewertung ökologisch wertvoller Lebensräume in der Nordsee. Final report for BfN (see: http://www.bfn.de/marinehabitate/de/downloads/berichte/Benthos_oekolog_Untersuchungen_Nordsee_2004.pdf) Bremerhaven

Rees H, Cochrane S, Craeymeersch J, de Kluijver M, Degraer S, Desroy N, Dewarumez JM, Duineveld G, Essink K, Hillewaert H, Kilbridge R, Kröncke I, Nehmer P, Rachor E, Reiss H, Robertson M, Rumohr H, Vanden Berghe E, Van Hoey G (2002) The North Sea benthos project: planning, management and objectives. ICES CM 2002/L:09, 1–10

Salm R, Price A (1995) Selection of marine protected areas. In: Gubbay S (ed) Marine protected areas – Principles and techniques for management. Chapman & Hall, London, pp 15–31

Salzwedel H, Rachor E, Gerdes D (1985) Benthic macrofauna communities in the German Bight. Veröff Inst Meeresforsch Bremerhaven 20:199–267

Stripp K (1969) Die Assoziationen des Benthos in der Helgoländer Bucht. Veröff Inst Meeresforsch Bremerhaven 12:95–142

Thorson G (1957) Bottom communities (sublittoral or shallow shelf). Mem Geol Soc Amer 67:461–534

Ursin E (1952) A quantitative investigation of the echinoderm fauna of the central North Sea. Meddr Danm Fisk- og Havunders 2:1–204

Chapter 8

Benthic assessment of marine areas of particular ecological importance within the German Baltic Sea EEZ

Michael L. Zettler[a] and Fritz Gosselck[b]

[a] *Baltic Sea Research Institute, Rostock*
[b] *Institute for Applied Ecology, Broderstorf*

Abstract

The Habitats Directive is one of the main legal tools of the European Union to preserve biodiversity by maintaining and restoring natural habitats, and establishing a network of protected sites (NATURA 2000). One point of interest is the characterisation of marine habitats to localise the areas that fulfil the protection targets. Of the habitats listed in Annex I of the Habitats Directive, mainly *reefs* and *sandbanks* are relevant in the case of German Baltic offshore waters. Along a strong salinity gradient (5 to 25 psu), four planned offshore Marine Protected Areas (MPAs) within the German Exclusive Economic Zone (EEZ) of the Baltic Sea were thoroughly investigated in this study. More than 250 locations were analysed using a combination of standard sampling methods, underwater video technique, and measurement of abiotic factors (salinity, oxygen, sediment parameters). The areas of interest were the *Odra Bank* (*Oderbank*), *Adler Ground* (*Adlergrund*), *Kadet Trench* (*Kadetrinne*), and *Fehmarn Belt* (*Fehmarnbelt*). The characteristic living communities (macrophytes, such as algae and sea grass, and macrozoobenthos such as worms, bivalves and crustaceans) for the different habitat types in these areas were characterised. Due to different salinity regimes, the benthic colonisation is different as well. In the Baltic Sea, with its decreasing salinity from west to east, the number of marine species declines, too. In the present study, altogether approximately 350 macrozoobenthic species and approximately 20 macrophytes were identified.

1 Introduction

During the last 5 years, extensive studies of the benthic habitat[1] within the proposed marine protected areas of German Baltic offshore waters were done by the authors (IfAÖ 1998, Zettler et al. 2003a, 2003b). In the present paper, four areas are relevant. These are the *Odra Bank* (which belongs to the *Pommeranian Bay* (*Pommersche Bucht*)), the *Adler Ground*, the *Kadet Trench* and the slope region of the *Fehmarn Belt* (figure 1).

In the following, we want to introduce these four important marine areas particularly with regard to benthic habitat characteristics and its colonisation by macrozoobenthos.[2] In terms of benthic biodiversity, it is taken into account that the German Baltic has a strong salinity gradient from the west (approx. 20–25 psu) to the east (app. 5–8 psu) in near-bottom water layers. Due to this gradient, the respective species distribution is limited. The mean species number decreases from about 700 in the *Kiel Bight* (*Kieler Bucht*, Gerlach 2000) to 360 in the *Mecklenburg Bay* (*Mecklenburger Bucht*), to only 50 in the *Pommeranian Bay* (Zettler and Röhner 2004).

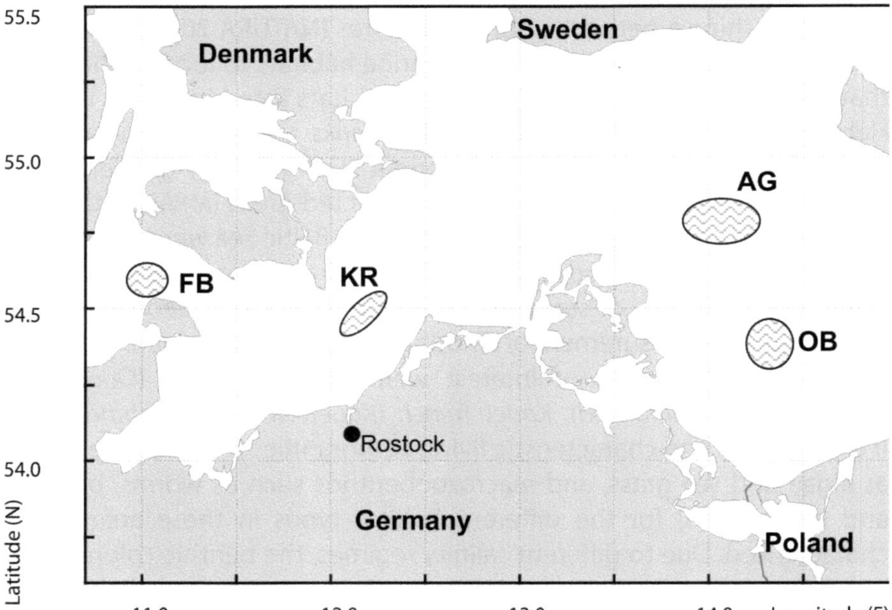

Figure 1. Investigation areas within German Baltic waters (The four areas of interest are indicated by dashed circles (OB=*Odra Bank*, AG=*Adler Ground*, KR=*Kadet Trench*, FB=*Fehmarn Belt*))

[1] Benthic habitat: habitat of organisms living in/on the seafloor.

[2] Macrozoobenthos: larger seafloor animals.

2 Methods

The data sets analysed were based on about 250 stations sampled by IfAÖ (1998) and Zettler et al. (2003a, 2003b, and unpublished) during the last 6 years. Benthic samples were taken with a 0.1m^2 van Veen grab. Due to sediment conditions, grabs of different weights were used. Three (or two) replicates of grab samples were carried out at each station. Additionally, a dredge haul (net mesh size 5 mm) was taken in order to obtain mobile or rare species. All samples were sieved through a 1-mm screen and animals were preserved in the field with 4% formaldehyde. For sorting in the laboratory, a stereomicroscope with 10–40x magnification was used. For the characterisation of the habitat (i.e., assessment of sediment structure, current and epibenthos[3]), an underwater video system mounted on a sledge was used.

3 Odra Bank

The *Pommeranian Bay* is a large shallow water area in the southwestern Baltic Sea with the *Odra Bank* located in its central part. The *Odra Bank*, a permanently submerged sandbank (in terms of the definition of the EC interpretation manual, European Commission 2003), with an average depth of 15 metres, rises up to 7–8 metres beneath the surface from the ground of the *Pommeranian Bay*. The north–south extension is about 35 km, and 25 km from the eastern to the western edge. Sediments are mainly constituted by fine sands enriched with a significant amount of shell gravel. Salinity is between 7–8 psu. The *Odra Bank* is assumed to be a submerged dune complex.

3.1 Macrophytes

The development of sea grass beds and the settlement of algae are prevented by substrate dynamics and limited penetration of light. However, drifting algae (e.g., *Cladophora glomerata*, *Ceramium diaphanum*, *C. strictum*, *Polysiphonia violacea*) and loose "sea grass" leaves (*Zostera marina*, *Zannichellia palustris*) can often be found. The coverage with drifting algae varies seasonally and inter-annually and can be significant.

3.2 Macrozoobenthos

The *Odra Bank* is habitat for a typical benthic community of limited species abundance where crustaceans, molluscs and polychaetes predominate.

[3] Epibenthos: organisms living on the seafloor surface.

Only a few species (for example, the bristleworm *Pygospio elegans*, the amphipod *Bathyporeia pilosa*, or the brown shrimp *Crangon crangon*) can cope with the extreme environmental conditions (high exposure, low salinity) and at the same time develop in high densities. Altogether, 21 species were recorded in this area (out of approximately 50 species known in the *Pommeranian Bay*, based on our own observations). Among the 21 species recorded were 4 polychaetes, 3 oligochaetes, and 6 crustaceans. However, only 2 out of the 6 species of crustaceans typically inhabit sandy substrates (*Bathyporeia pilosa*, and *Crangon crangon*). The others can be found among drifting algae and *Mytilus* aggregations. There were also several species of molluscs, the laver spire shell *Hydrobia ulvae*, and 4 bivalves species to be found (blue mussel *Mytilus edulis*, lagoon cockle *Cerastoderma glaucum*, soft-shell clam *Mya arenaria* and Baltic tellin *Macoma balthica*). Hydrozoans, bryozoans, as well as nemertineans were recorded.

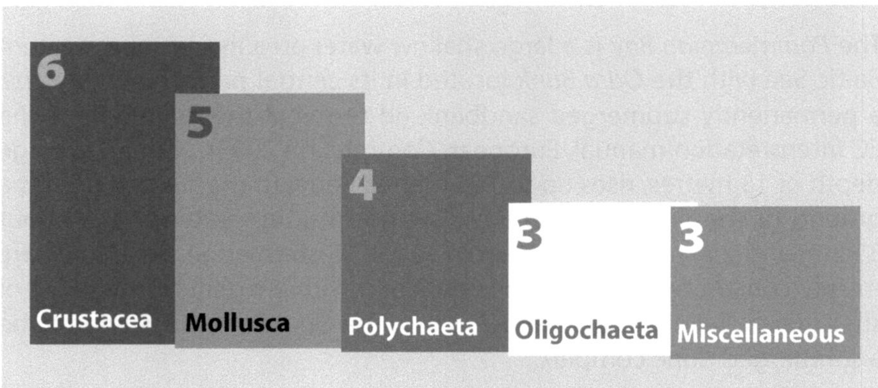

Figure 2. Species composition at the *Odra Bank* area in 1998 (Altogether 21 taxa were found.)

Although species abundance is low in the *Odra Bank* area, sediments are densely populated. Mean density was above 5,000 individuals per square metre (ind./m^2) (table 1). Only 3 species, *Bathyporeia pilosa*, *Mya arenaria* and *Hydrobia ulvae*, counted for 72% of all individuals. Densities of 200–400 ind./m^2 were observed for the bristleworm *Pygospio elegans* and the ragworm *Hediste diversicolor*, and the bivalves *Cerastoderma glaucum* and *Macoma balthica*. Other species populated the area with less than 100 ind./m^2. Drifting *Mytilus* aggregations occurred in varying densities. Therefore, density of the fauna associated with these aggregations also varied considerably. Both the gammarid shrimps *Gammarus salinus* and *G. zaddachi* and the isopods *Jaera albifrons* and *Idotea balthica* are especially linked to these structures.

Table 1. Dominating species at *Odra Bank* in 1998 (Sandy sediments are characterised by endobenthic species[4]. Epibenthic species (excepting *Mytilus edulis*) are not taken into account. The presence is the percentage of sampling stations where the species was recorded.)

Category	Taxon	Common name	Presence	ind./m^2
AMPHIPODA	*Bathyporeia pilosa*	Sand digger shrimp	100%	1,780
BIVALVIA	*Mya arenaria*	Soft-shell clam	94%	1,436
GASTROPODA	*Hydrobia ulvae*	Laver spire shell	82%	594
POLYCHAETA	*Pygospio elegans*	Bristleworm	94%	393
BIVALVIA	*Cerastoderma glaucum*	Lagoon cockle	100%	376
BIVALVIA	*Macoma balthica*	Baltic tellin	100%	309
POLYCHAETA	*Hediste diversicolor*	Ragworm	100%	188
BIVALVIA	*Mytilus edulis*	Blue mussel	100%	85
POLYCHAETA	*Marenzelleria neglecta*	Bristleworm	94%	70
OLIGOCHAETA	*Tubifex costatus*	Sludge-worm	88%	42

The distribution of species that typically inhabit sandy substrates was homogenous. Their presence in the area was high. Ten (10) species were recorded at more than 80% of all stations. The bristleworm *Streblospio dekhuyzeni* and the Enchytraeidae (Oligochaeta) were found at more than 50% of all stations. Apart from the Nemertini and another species of Oligochaeta, remaining taxa can be regarded as fauna associated with drifting algae and *Mytilus* aggregations; these occurrences depend on abiotic conditions (e.g., current, sand ripple marks).

3.3 Ecological evaluation of the Odra Bank area

Sandbanks are feeding grounds for wintering and moulting seabirds, and of various species of fish. Blue mussel (*Mytilus edulis*) and soft-shell clam (*Mya arenaria*) are the main prey for black scooters (*Melanitta nigra*) and long-tailed ducks (*Clangula hyemalis*) (Kube 1996). The sea ducks prefer the shallow sandbanks because of the low-diving depths and they feed on the plentiful juvenile and thin-shelled bivalves of the blue mussel and soft-shell clam.

[4] Endobenthic species: species living in bottom sediments.

4 Adler Ground

The *Arkona Basin* (*Arkona Becken*) is regarded as a part of the transitional zone between the Kattegat and the deep basins of the Baltic proper. It covers an area of about 19,000 km^2. About a quarter of this area is deeper than 40 metres (maximum 53 metres). Only the marginal zones (like the *Adler Ground*) are shallower. The *Adler Ground* is a glacial-morphological structure and is a part of the Rönne Bank system. It is situated in the southeastern part of the *Arkona Basin*. It divides this Basin from the more easterly-lying *Bornholm Basin* (*Bornholm Becken*). In 2002 and 2003, the authors investigated the western part of this area between depths of 5 and 45 metres. The *Adler Ground* is a rise from the deepest muddy basin (45 metres) to the shallow stony and boulder grounds (up to 5 metres). The slope with sandy sediments and many bigger stone fields with moderate salinity conditions were especially of interest. In terms of the interpretation manual of the European Commission (2003), it is in and among reefs, submarine rocky substrates, and biogenic concretions (blue mussel banks, cold water coral reefs) – which arise from the seafloor in the sublittoral[5] zone but may extend into the littoral zone[6] – where there is an uninterrupted zonation of plant and animal communities. These reefs generally support a zonation of benthic communities of algae and animals species including concretions and encrustations.

4.1 Macrophytes

Down to a 12-metre depth, a more or less dense coverage of brown algae was observed. The bottom consists mainly in bigger stones, boulders, gravel and sand. The algae, principally formed by *Fucus serratus*, are fixed on stony substrates. *Laminaria saccharina* settled in very low numbers, too. Further, some red algae were also observed. Filamentous and drifting algae (*Polysiphonia* spp. and *Ahnfeltia plicata*) settle in depths of up to 20 metres. In some areas (especially between 6- and 11-metre water depth), the seaweeds of *Chorda tomentosa* are noteworthy.

4.2 Macrozoobenthos

The benthic community at the *Adler Ground* depends on the water depth and sediment characteristics. Deeper than 35 metres, the sediment is

---

5 Sublittoral zone: zone between the low-tide mark and the edge of the continental shelf.

6 Littoral zone: zone between the tide marks (also intertidal zone). The Baltic Sea does not have tide; here the littoral zone extends from the shoreline to a few metres deep.

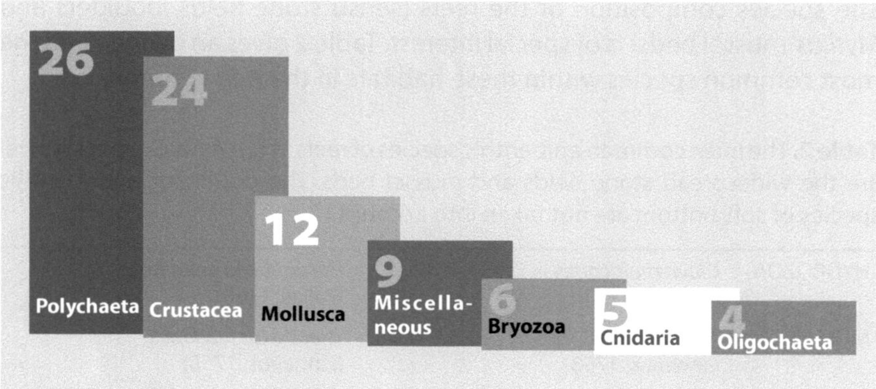

Figure 3. Species composition at the *Adler Ground* area in 2002/2003 (Altogether 86 taxa were found.)

muddy, the temperature is low, and the salinity is quite high in comparison with shallower zones. The salinity regime ranges between 15 and 20 psu. Due to saline submergence,[7] some marine species are able to settle in these deep waters. Some glacial relict species (e.g., the amphipod *Monoporeia affinis*), boreal species (the amphipod *Pontoporeia femorata*, or the bivalves *Astarte borealis*, *A. elliptica*, *Arctica islandica*) and marine species (e.g., the polychaete worms *Harmothoe imbricata*, *Ampharete baltica*, *Phyllodoce mucosa*) are typical for this muddy zone. At the slope and on the top of the rise, completely different substrates (sand, stones, and boulders) and salinity conditions (7–10 psu) were observed. Due to these conditions, a clear and abrupt change in the species community was observed. Endobenthic species dominate the sandy zones between the stones and boulders. The bivalves *Macoma balthica* and *Mya arenaria*, the polychaetes *Pygospio elegans* and *Scoloplos armiger* and the crustaceans *Bathyporeia pilosa* and *Corophium volutator* settled here. Within the stony areas, more or less epibenthic species dominate the community. Here belong the hydrozoan *Clava multicornis*, the cirripedian *Balanus improvisus* (in deeper and more saline waters, *B. crenatus*), the amphipod *Gammarus* spp., the blue mussel *Mytilus edulis*, and the bryozoan *Electra crustulenta*.

[7] Saline submergence: more saline water has a higher density, is heavier than the surrounding water and sinks down to the bottom.

The species composition of the reefs (sensu stone fields, boulders and *Mytilus*-mussel beds) is of special interest. Table 2 gives an overview of the most common species within these habitats in the *Adler Ground*.

Table 2. The most common epibenthic species of reefs at the *Adler Ground* (Typical are the widespread stone fields and mussel beds. The dominant endobenthic species of soft bottom are not taken into account.)

HYDROZOA	*Clava multicornis* (Forskal, 1775)	*Hartlaubella gelatinosa* (Pallas, 1766)
MOLLUSCA	*Mytilus edulis* Linnaeus, 1758	*Theodoxus fluviatilis* (Linnaeus, 1758)
CRUSTACEA	*Balanus improvisus* Darwin, 1854	*Jaera albifrons* Leach, 1814
	Saduria entomon (Linnaeus, 1758)	*Gammarus oceanicus* Segerstrale, 1947
	Gammarus salinus Spooner, 1947	*Gammarus zaddachi* Sexton, 1912
	Palaemon elegans (Linnaeus, 1758)	*Rhithropanopeus harrisii* (Gould, 1841)
BRYOZOA	*Callopora lineata* (Linnaeus, 1767)	*Electra crustulenta* (Pallas, 1766)

4.3 Ecological evaluation of the Adler Ground area

From the macrozoobenthic point of view, the stone fields, boulder grounds, and mussel beds are particularly valuable. These epibenthic structures form suitable habitat conditions for many species. The three-dimensional living space is compartmentalised into countless small holes and caves. Due to these opportunities, the benthic biodiversity in this area increased in comparison with more homogeneous sandy or muddy substrates. These stone fields, boulder grounds and mussel beds are in shallow waters and go to a depth of 30 metres. Due to saline submergence, higher saline conditions were found in this deeper zone of the slope. The combination of waters that are more saline and optimal substrates allow some marine species to settle here. The *Adler Ground* is a conglomerate of many rocky reefs and mussel beds and it builds an island of high biodiversity in a region of depleted invertebrate community. About 90 macrozoobenthic species live in the *Adler Ground* area. Twelve species belong to the German red list (Gosselck et al. 1996, Zettler et al. 2003a, 2003b).

5 Kadet Trench

The *Kadet Trench* lies in the transition zone between the *Mecklenburg Bay* and the *Arkona Basin*. The depth ranges between 15 and 32 metres. The mean salinity at the bottom varies from 11 to 20 psu. The strait is the main gateway for Baltic water exchange. It connects two basins where fine-grained sediments accumulate, i.e., the *Mecklenburg Bay* to the southwest and the *Arkona Basin* to the northeast. Owing to the indrift of larvae, the diversity of the fauna is particularly high here. The strong current prevents the siltation of the deep channel. Typical substrates are residual sediments like blocks, stones and boulders exposed in the bottom. In some years, though rarely, oxygen-depleted water from the *Kiel Bight* penetrates into the *Kadet Trench*. After these strong events, a mass mortality of most macrozoobenthic taxa could be observed. Only a few species, e.g., bigger bivalves (*Arctica islandica, Astarte* spp.) could survive. Under suitable conditions, the community is regenerated only few months later, consisting of the same species composition (though not the same age structure since the community lacks older individuals.)

5.1 Macrophytes

Only a few macrophytes find suitable living conditions in the *Kadet Trench*. The rocky substrates are in fact good for the settlement of brown algae, but due to the quite deep range (25–30 metres), there are insufficient light conditions. The sugar kelp *Laminaria saccharina* was observed only in some parts of the slope. The sea beech *Delesseria sanguinea* (very sparse) finds here its eastern-most distribution within the Baltic.

5.2 Macrozoobenthos

Within the *Kadet Trench*, the benthic community varies as a consequence of patchy substrate conditions. Rocky substrates (stones, boulders, gravel) dominate at the bottom and at the slope of the channel. Therefore, a hard-bottom community is characteristic for this habitat. Typical species are sponges (e.g., *Halichondria panicea*), the anemone *Metridium senile*, the blue mussel *Mytilus edulis* and ascidians (e.g., *Dendrodoa grossularia*) (see table 3).

Table 3. The most common epibenthic species of reefs at the *Kadet Trench* area (The widespread stone fields and boulders are typical. The dominant endobenthic species of soft bottom are not taken into account.)

PORIFERA	*Halichondria panicea* (Pallas, 1766)	*Haliclona limbata* (Pallas, 1766)
CNIDARIA	*Metridium senile* (Linnaeus, 1761)	*Opercularella lacerata* (Johnston,1847)
MOLLUSCA	*Hiatella arctica* (Linnaeus, 1767)	*Mytilus edulis* Linnaeus, 1758
	Acanthodoris pilosa (Abildgaard, 1789)	*Retusa truncatula* (Bruguiere, 1792)
POLYCHAETA	*Nepthys caeca* (Fabricius, 1780)	*Nereimyra punctata* (O.F. Müller, 1788)
	Phyllodoce mucosa Oersted, 1843	*Polydora ciliata* (Johnston, 1838)
CRUSTACEA	*Balanus crenatus* Bruguiere, 1789	*Idotea balthica* (Pallas, 1772)
	Corophium crassicorne Bruzelius, 1859	*Gammarus oceanicus* Segerstrale, 1947
	Gammarus salinus Spooner, 1947	*Gammarus zaddachi* Sexton, 1912
PYCNOGONIDA	*Nymphon brevirostre* Hodge, 1863	
BRYOZOA	*Callopora lineata* (Linnaeus, 1767)	*Electra pilosa* (Linnaeus, 1767)
	Eucratea loricata (Linnaeus, 1758)	*Valkeria uva* (Linnaeus, 1758)
ECHINODERMATA	*Asterias rubens* Linnaeus, 1758	*Dendrodoa grossularia* (Van Beneden, 1846)

In spaces between bigger stones and in areas of fewer current, soft substrates accumulate. In these zones, endobenthic species reach quite high abundance. Due to the good salinity conditions (11–20 psu), the *Kadet Trench* constitutes for many marine species their eastern-most distribution. Typical species living at the edge of their range are the bivalves *Abra alba*, *Musculus discors*, *Mysella bidentata* and *Hiatella arctica*; the polychaetes *Lepidonotus squamatus*, *Nereimyra punctata*, *Pherusa plumosa* and *Scalibregma inflatum*; and the crustaceans *Balanus crenatus* and *Dyopedos monacanthus*. Further, some "exotic species" like *Nymphon brevirostre* and *Ophiura albida* were found. With regard to presence and

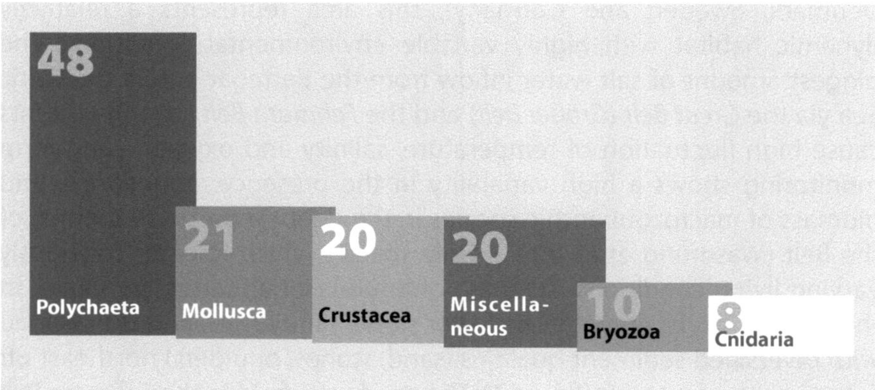

Figure 4. Species composition at the *Kadet Trench* area in 2000 (Altogether 127 taxa were found.)

abundance, the bivalve *Macoma balthica* and the polychaetes *Bylgides sarsi*, *Scoloplos armiger* and *Heteromastus filiformis* are the most dominant species. Further, the ocean quahog (*Arctica islandica*) and the cumacean *Diastylis rathkei* reach quite high abundance and biomass. Altogether, 127 species were found in 2000 (figure 4).

The dominant groups are the polychaetes, followed by molluscs and crustaceans. From the literature and own studies, about 170 species for the *Kadet Trench* area are known with similar relations (Zettler and Röhner 2004).

5.3 Ecological evaluation of the Kadet Trench area

From the macrozoobenthic point of view and with respect to conservation strategies, the areas of the slope and furrow are especially valuable. These zones are characterised by aggregates of stones and boulders with a rich epibenthic fauna. The combination of soft bottoms and rocky substrates and the moderate salinity regime allow a high benthic biodiversity. Typically, many marine species find its eastern distribution limit within the *Kadet Trench*. In 2000, altogether 127 species were found (Zettler unpublished). Twenty-three belong to the German red list (Gosselck et al. 1996).

6 Fehmarn Belt

The *Fehmarn Belt* is the connection between the *Kiel Bight* in the west and the *Mecklenburg Bay* in the east. As part of the Belt Sea (area between

Denmark, Sweden and Germany), this area represents a relatively dynamic habitat with highly variable environmental conditions. The biggest amount of salt water inflow from the Kattegat enters the Baltic Sea via the *Great Belt* (*Großer Belt*) and the *Fehmarn Belt*. Strong currents cause high fluctuation of temperature, salinity and oxygen. Long-term monitoring shows a high variability in the presence, abundance, and biomass of macrozoobenthic species in the deepest parts (30 metres) of the Belt (Wasmund et al. 2003). Only species which tolerate constantly varying living conditions (such as continuous inhabitants) are found. In the present study, only the shallower slope, sandy rises, and the plateau with diversified sediment qualities (sand, stones, boulders) northeast off Fehmarn are of interest (figure 1). The sandwave field in the *Fehmarn Belt* is generated by intermittent currents. Due to the good salinity condition (15–23 psu) caused by salt-rich water from the *Great Belt*, and the stable oxygen supply in the medium water depth zone (15–25 metres), many marine and euryhaline species (e.g., the common whelk *Buccinum undatum*, the hermit crab *Pagurus bernhardus*, the green sea urchin *Psammechinus miliaris*) find suitable living conditions here.

6.1 Macrophytes

Widespread colonies of bigger algae like *Laminaria saccharina* and *Delesseria sanguinea* (figure 5) are found especially on stable sandfields and on stones and boulder areas. In some regions, these plants cover about 80% of the surface. The rocky substrates build suitable settlement conditions for the rhizomes. The higher plant *Zostera marina* was found sparsely in the shallower sandy zones. Further, many species of drifting red algae and some filamentous green algae were observed.

Figure 5. Very common on bigger stones in water depths beween 15 and 25 meters in the *Fehmarn Belt* area is the sea beech *Delesseria sanguinea*

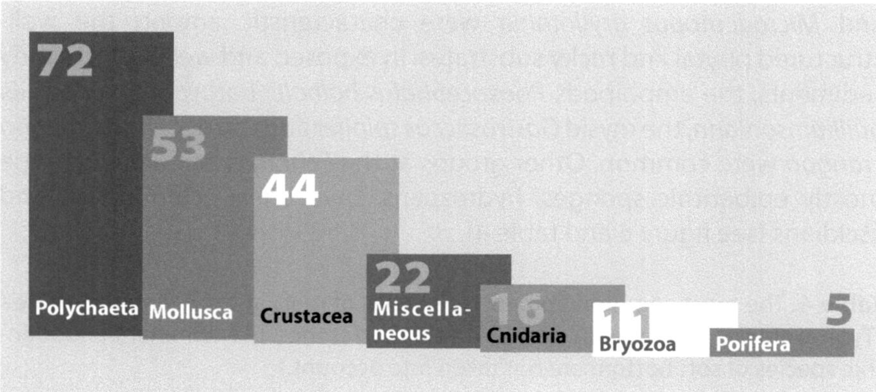

Figure 6. Species composition at the *Fehmarn Belt* area in 2003 (Altogether 223 taxa were found.)

6.2 Macrozoobenthos

During the investigations in 2003, the *Fehmarn Belt* proved to be highly biodiversed. About 223 taxa were observed (figure 6). The polychaetes were the largest category, with 72 species. In the soft bottom, the endobenthic *Ampharete baltica, Aricidea suecica, Heteromastus filiformis, Polydora quadrilobata, Scoloplos armiger* and *Spio goniocephala* showed highest presence and abundance. More linked to rocky substrates or plants, the bristleworms *Flabelligera affinis, Lepidonotus squamatus, Eulalia bilineata* and *Nereimyra punctata* were found. The second largest category were the molluscs. Endobenthic bivalves dominated (mainly the biomass) in sandy substrates. Some examples are *Abra alba, Arctica islandica, Astarte borealis, Corbula gibba, Macoma calcarea, Mya arenaria* and *Parvicardium ovale.* Hard bottom or plants were frequently colonised by epibenthic bivalves and gastropods like *Buccinum undatum, Facelina bostoniensis, Hiatella arctica, Lacuna pallidula, Mytilus edulis, Modiolus modiolus, Onoba semicostata* and *Retusa truncatula.* Within sponges and ascidians, bivalves of the genus *Musculus* (*M. discors* and *M. marmoratus*) were ingrown. In addition, the introduced North American bivalve species *Ensis americanus* has stable populations in the coarse-sand sediments here. In the *Fehmarn Belt* area, the species lives at the margin of its salinity tolerance.

The third largest category were the crustaceans (figure 6). Forty-four (44) species were verified for the *Fehmarn Belt* area. Most of them are vagile (i.e., are able to move about) and epibenthic. With 25 species, the amphipods composed the main part of this group. *Apherusa bispinosa, Caprella septemtrionalis, Corophium insidiosum, Gammarellus homari* and *Microdeutopus gryllotalpa* were characteristic among the well-structured phytal and rocky substrates. In exposed and well-sorted sandy sediments, the amphipods *Phoxocephalus holbolli, Bathyporeia pilosa, B. guilliamsoniana,* the mysid *Gastrosaccus spinifer* and the decapod *Crangon crangon* were common. Other groups at the *Fehmarn Belt* area were the mostly epibenthic sponges, hydrozoans, bryozoans, echinoderms and ascidians (see figure 6 and table 4).

Table 4. The most common epibenthic species of reefs at the *Fehmarn Belt* area (Typical are the widespread *Laminaria* fields and stones. The dominant endobenthic species of soft bottom are not taken into account.)

PORIFERA	*Halichondria panicea* (Pallas, 1766)	*Haliclona limbata* (Pallas, 1766)
	Halisarca dujardini Johnston, 1842	*Leucosolenia* sp.
CNIDARIA	*Metridium senile* (Linnaeus, 1761)	*Hartlaubella gelatinosa* (Pallas, 1766)
	Opercularella lacerata (Johnston, 1847)	*Sertularia cupressina* Linnaeus, 1758
MOLLUSCA	*Hiatella arctica* (Linnaeus, 1767)	*Musculus discors* (Linnaeus, 1767)
	Musculus marmoratus (Forbes, 1838)	*Acanthodoris pilosa* Abildgaard, 1789)
	Amauropsis islandica (Gmelin, 1791)	*Bittium reticulatum* (Da Costa, 1778)
	Buccinum undatum (Linnaeus, 1758)	*Facellina bostoniensis* (Couthouy, 1838)
	Lacuna pallidula (Da Costa, 1779)	*Neptunea antiqua* (Linnaeus, 1758)
	Onoba semicostata (Montagu, 1803)	*Retusa truncatula* (Bruguiere, 1792)
POLYCHAETA	*Flabelligera affinis* M. Sars, 1829	*Harmothoe imbricata* (Linnaeus, 1767)
	Harmothoe impar (Johnston, 1839)	*Lepidonotus squamatus* (Linnaeus, 1758)
	Nepthys caeca (Fabricius, 1780)	*Microphthalmus aberrans* (Linnaeus, 1758)
	Nereimyra punctata (O.F. Müller, 1788)	*Nereis pelagica* Linnaeus, 1758
	Nicolea zostericola (Oersted, 1844)	*Phyllodoce mucosa* Oersted, 1843
	Polydora ciliata (Johnston, 1838)	*Streptosyllis websteri* Southern, 1914

Table 4. continued

CRUSTACEA	*Balanus crenatus* Bruguiere, 1789	*Praunus inermis* (Rathke, 1843)
	Idotea balthica (Pallas, 1772)	*Idotea granulosa* Rathke, 1843
	Apherusa bispinosa (Bate, 1856)	*Caprella linearis* (Linnaeus, 1767)
	Caprella sepentrionalis Kröyer, 1838	*Cheirocratus sundevalli* (Rathke, 1843)
	Corophium crassicorne Bruzelius, 1859	*Dexamine spinosa* (Montagu, 1813)
	Gammarellus homari (Fabricius, 1779)	*Microdeutopus gryllotalpa* Da Costa, 1853
	Phtisica marina Slabber, 1769	*Carcinus maenas* (Linnaeus, 1758)
PYCNOGONIDA	*Callipalene brevirostris* (Johnston, 1837)	*Nymphon brevirostre* Hodge, 1863
BRYOZOA	*Callopora lineata* (Linnaeus, 1767)	*Cribrilina punctata* (Hassall, 1841)
	Electra pilosa (Linnaeus, 1767)	*Eucratea loricata* (Linnaeus, 1758)
	Flustra foliacea Linnaeus, 1758	*Mucronella immersa* (Fleming, 1882)
ECHINODERMATA	*Asterias rubens* Linnaeus, 1758	*Psammechinus miliaris* (P.L.S. Müller, 1766)
ASCIDIACEA	*Ciona intestinalis* (Linnaeus, 1767)	*Dendrodoa grossularia* (Van Beneden, 1846)

6.3 Ecological evaluation of the Fehmarn Belt area

Due to the good salinity conditions (15–23 psu), the *Fehmarn Belt* area constitutes for many marine species the only distribution in the Baltic Sea area. Literature data (e.g., Kock 2001, Zettler et al. 2000, and in preparation) and own results show that the *Fehmarn Belt* area has a potential macrozoobenthic inventory of about 300 species. The *Fehmarn Belt* area builds a refuge for highly biodiversed macrozoobenthic and macrophytobenthic communities. Several species (approx. 40) belong to the red list. The rocky habitats and the field of red and brown algae especially form suitable living conditions for many sensitive species. The closeness of the *Great Belt* and its inflow from the Kattegat area cause a relatively regular supply of salt-rich water and oxygen.

Acknowledgements

The study was financed mostly by the German Federal Agency of Nature Conservation (BfN).

References

European Commission (2003) DG Environment Nature and biodiversity, Interpretation Manual of European Union Habitats. Version EUR 25 April 2003

Gerlach A (2000) Checkliste der Fauna der Kieler Bucht und eine Bibliographie zur Biologie und Ökologie der Kieler Bucht. In: Bundesanstalt für Gewässerkunde (ed) Die Biodiversität in der deutschen Nord- und Ostsee. Band 1. Bericht BfG-1247, Koblenz, pp 1–376

Gosselck F, Arlt G, Bick A, Bönsch R, Kube J, Schroeren V, Voss J (1996) Rote Liste und Artenliste der benthischen wirbellosen Tiere des deutschen Meeres- und Küstenbereichs der Ostsee. Schr-R Landschaftspfl Natursch 48:41–51

IfAÖ (1998) Wissenschaftliche Grundlagen zur Ausweisung und zum Management mariner off-shore-Schutzgebiete im Bereich der Hoheitsgewässer und der Ausschließlichen Wirtschaftszone Deutschlands in der Ostsee und deren Integration in das System von Baltic Sea Protected Areas (BSPAs). Gutachten für das Bundesamt für Naturschutz (Vilm)

Kock M (2001) Untersuchungen des Makrozoobenthos im Fehmarnbelt, einem hydrographisch besonders instabilen Übergangsbereich zwischen zentraler und westlicher Ostsee. Dissertation Universität Kiel

Kube J (1996) Spatial and temporal variations in the population structure of the soft-shell clam *Mya arenaria* in the Pomeranian Bay (Southern Baltic Sea). J Sea Research 35:335–344

Wasmund N, Pollehne F, Postel L, Siegel H, Zettler ML (2003) Biologische Zustandseinschätzung der Ostsee im Jahre 2002. Meereswissenschaftliche Berichte 56:1–78

Zettler ML, Röhner M (2004) Verbreitung und Entwicklung des Makrozoobenthos der Ostsee zwischen Fehmarnbelt und Usedom - Daten von 1839 bis 2001. In: Bundesanstalt für Gewässerkunde (ed) Die Biodiversität in Nord- und Ostsee. Band 3. Bericht BfG-1421, Koblenz, pp 1–175

Zettler ML, Bönsch R, Gosselck F (2000) Verbreitung des Makrozoobenthos in der Mecklenburger Bucht (südliche Ostsee) – rezent und im historischen Vergleich. Meereswissenschaftliche Berichte 42:1–144

Zettler ML, Röhner M, Frankowski J, Becher H, Glockzin I (2003a) F+E-Vorhaben „Benthologische Arbeiten zur ökologischen Bewertung von Windenergie-Anlagen-Eignungsgebieten in der Ostsee" – Endbericht für die Areale Kriegers Flak (KF) und Westlicher Adlergrund (WAG). Forschungsbericht im Auftrag des BfN

Zettler ML, Röhner M, Frankowski J, Glockzin I (2003b) Benthologische Arbeiten zur ökologischen Bewertung von Windenergie-Anlagen-Eignungsgebieten in der Ostsee. Informationen über Lebensraumtypen in potentiellen NATURA 2000-Gebieten gemäß FFH-RL und deren benthische Besiedlung – Endbericht für die Areale Westliche Rönnebank (WRB), Kadetrinne (KR) und Fehmarnbelt (FB)

Part IV: Identification and assessments of sites – fish, mammals and birds

Chapter 9

Survey of NATURA 2000 fish species in the German North and Baltic Seas

Ralf Thiel[a] and Ina Backhausen[b]

[a] *German Oceanographic Museum, Stralsund*
[b] *German Federal Agency for Nature Conservation, branch office: Isle of Vilm*

Abstract

Due to their limited salinity tolerance, the occurrence of freshwater Annex II-fish species such as asp (*Aspius aspius*), spined loach (*Cobitis taenia*), bullhead (*Cottus gobio*), weatherfish (*Misgurnus fossilis*), bitterling (*Rhodeus amarus*), and ziege (*Pelecus cultratus*) was found to be restricted to freshwater and oligohaline-mesohaline habitats (i.e., areas with relatively low salinities), especially in the coastal waters of the German Baltic areas.

Mainly anadromous Annex II-fish species such as sturgeon (*Acipencer* spec.), North Sea houting (*Coregonus oxyrinchus*), river lamprey (*Lampetra fluviatilis*), sea lamprey (*Petromyzon marinus*), allis shad (*Alosa alosa*), and twaite shad (*Alosa fallax*) were expected to occur in the German parts of the North Sea and Baltic Sea. However, *Acipenser sturio*, a species of sturgeon, is considered extinct as a reproductive species in the North Sea region, and the sturgeon species last recorded in individual catches in the Baltic Sea in the 1990s is *A. oxyrinchus*.

At present, the North Sea houting is very rare in the North Sea, while the Baltic houting population (*Coregonus maraena*) is not protected under the Habitats Directive.

Since 1978, a total of 178 records of lampreys have been obtained in the German North Sea areas. Most of the lamprey records (mainly river lampreys) were located in nearshore areas, especially between

Helgoland Island and the mouth of the Elbe and Weser estuaries. Recently, 81 individuals of river lamprey and 4 individuals of sea lamprey were recorded in German Baltic waters from 2000 to 2004.

Today the main distribution range of allis shad is restricted to the Atlantic coasts of France and Portugal. The record of allis shad in the *Strelasund* in 1998 was the only specimen of this species caught in German Baltic waters during the last 20 years. Therefore, the species is expected to occur only accidentally in the German waters of the North and Baltic Seas.

In contrast, within the German coastal waters (12-nautical mile zone) of the North Sea, there were high-density areas of twaite shad. In total, 6,052 individuals of twaite shad have been caught in the *German Bight* (*Deutsche Bucht*) since 1978. Since 1995, however, the catch effort and the percentage of stations with twaite shad records have increased and, in general, there have been more records of this species in the German coastal waters than in the German Exclusive Economic Zone (EEZ).

From May 2003 until November 2004, 38 individuals of twaite shad were found at 14 different locations in the German EEZ of the Baltic Sea, northeast of the Rügen Island, as well as in the *Pommeranian Bay* (*Pommersche Bucht*) and the *Szczecin Lagoon* (*Stettiner Haff*) and its adjacent waters. Fifty percent (50%) of the total twaite shad records originated from the *Szczecin Lagoon* and adjacent waters. All of them were adult individuals. In the *Odra Bank* (*Oderbank*) and in the coastal waters of the Usedom Island, only juvenile individuals of age group 0 were caught; this amounted to 19% of the total number of records. Thirty-one percent (31%) of all individuals of twaite shad (mainly adults) were recorded from the potential Site of Community Interest (pSCI) *Western Rønne Bank* (*Westliche Rönnebank*) and adjacent waters. Given the recent records from the German Baltic waters, it is assumed that, after about 50 years of decline, the Baltic population of twaite shad has been increasing since the middle of the 1990s.

The observed status of the populations of Annex II-fish species in the German waters of the Baltic Sea and North Sea indicate that the study of their distribution and the trends in their population development must be continued on the basis of an international cooperation, especially with the new EU Member States of the southern Baltic.

1 Introduction

In order to fulfil the requirements of the European Directive on the conservation of natural habitats and of wild fauna and flora (the "Habitats Directive" 92/43/EEC of 21 May 1992), it is necessary to investigate the occurrence and distribution – within the EEZ of Germany in the North- and Baltic Seas – of fish species listed in Annex II of the Directive. These listed species require the identification and designation of Special Areas of Conservation (SACs). In general, there is little knowledge on the importance of European marine and brackish waters as habitat for Annex II-fish species (Elliott and Hemingway 2002).

In 2002, the Federal Research Centre for Fisheries (BFAFi) and the Carl von Ossietzky University (ICBM) were commissioned by the Federal Agency for Nature Conservation (BfN) for a research and development project. The project's goal was to assess whether areas of high abundance of Annex II-fish species in the German EEZ can be identified or not. The ICBM could only analyse the distribution of twaite shad in the German North Sea because there was sufficient database for geo-statistical methods only for this species in this area (Stelzenmüller and Zauke 2003, Stelzenmüller et al. 2004). However, Kloppmann et al. (2003) demonstrated that there was an incomplete picture and database of the distribution of all Annex II-fish species in the German EEZ. In order to produce a sufficient dataset, the German Oceanographic Museum (DMM) and the University of Rostock (IfB) conducted investigations in 2003 in the Baltic Sea, especially in the *Pommeranian Bay* and adjacent waters. The aim of the study was to develop and test more specialised data enquiry and sampling strategies to find out if additional records of Annex II-fish species can be produced. The first results of this continuing study were published by Thiel et al. (2004a, 2004b, 2005).

2 Relevant Annex II-fish species and important aspects of their status

Only a few of the Annex II-fish species are expected to occur in the German parts of the North Sea and Baltic Sea (figure 1). These are mainly anadromous species (i.e., migrating from sea to freshwater to spawn) such as sturgeon, North Sea houting, river lamprey, sea lamprey, allis shad, and twaite shad. Moreover, some freshwater fish species listed in Annex II of the Habitats Directive may occur in the German Baltic Sea, especially in its eastern parts and nearshore areas with lowered salinities. Such species

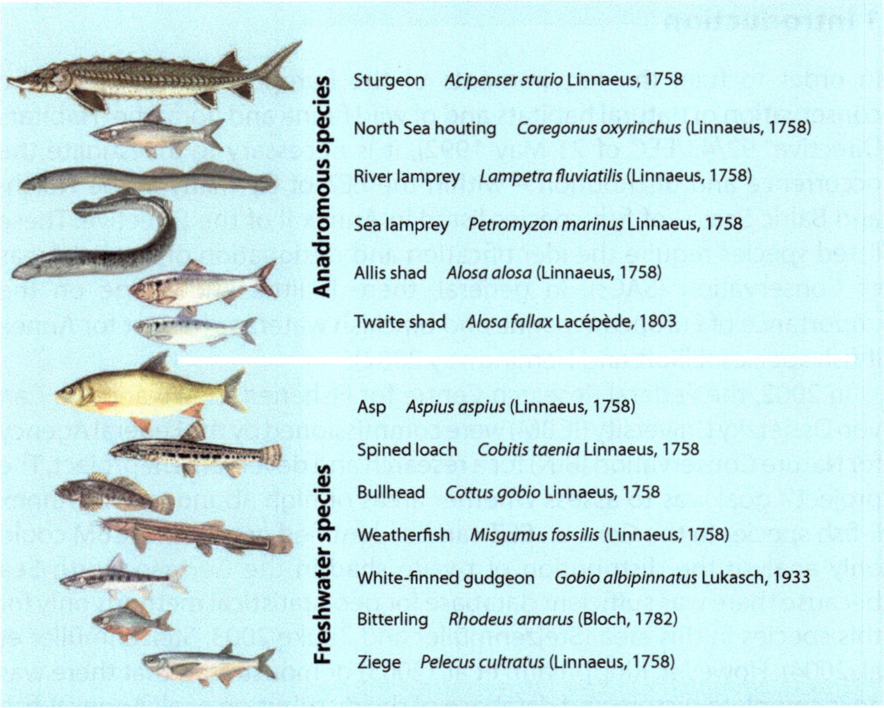

Figure 1. Relevant Annex II-fish species in the German North- and Baltic Seas

are asp, spined loach, bullhead, weatherfish, white-finned gudgeon, bitterling and ziege (figure 1).

Anadromous sturgeon populations were known from the North- and Baltic Seas and from most of the larger rivers draining into these marine waters (Freyhof 2002). According to Debus (1995), the extinction of these sturgeon populations during the last century was caused mainly by overfishing, pollution, river regulation, and damming in the North- and Baltic Sea areas (figure 2).

After 1950, sturgeons were only caught occasionally in the North- and Baltic Seas. *Acipenser sturio* (last caught in 1993) is considered extinct in Germany as a reproductive sturgeon species and is now reduced to a relict population in the French Gironde River (Kirschbaum and Gessner 2002). Recent research shows that the sturgeon species last recorded in individual catches in the Baltic in the 1990s was the Atlantic sturgeon *Acipenser oxyrinchus* Mitchill, 1815, a species still comparatively abundant along the northeastern coast of the United States (Ludwig et al. 2002).

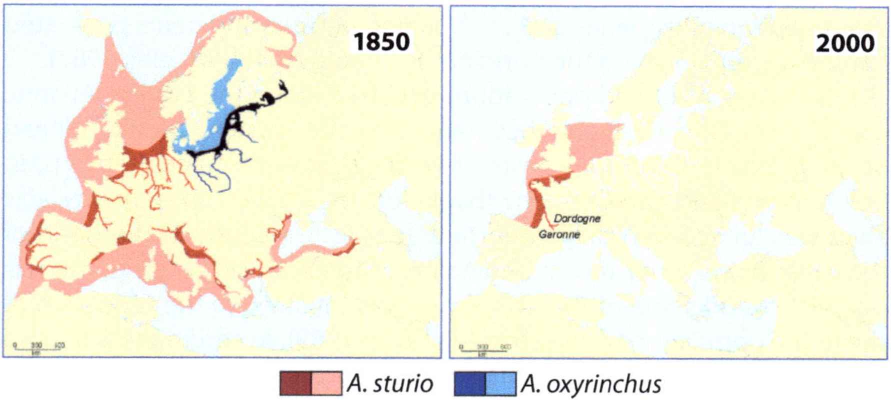

Figure 2. Distribution of European sturgeon (*A. sturio*) and Atlantic sturgeon (*A. oxyrinchus*) in Europe around 1850 (left) and 2000 (right)

Archaeological and genetic studies have shown that about one thousand years ago, in the Baltic, *A. oxyrinchus* replaced *A. sturio*, which until then had been the dominant native species. Because of this historic presence of *A. oxyrinchus* in the Baltic Sea, the re-introduction of this species would be justified and in line with respective legal guidelines.

The houting was a frequent anadromous fish in the coastal areas of the North Sea, especially in the Wadden Sea and in the large German North Sea estuaries (Vorberg and Breckling 1999). Nowadays, the houting is very rare in the North Sea. Reasons why it vanished from the German North Sea areas are probably the same as for the extinction of sturgeons (Freyhof 2002). However, the Baltic houting population was not protected under the Habitats Directive until now.

Both the river lamprey and sea lamprey occur in marine and brackish waters of the German North Sea and Baltic Sea region (Diercking and Wehrmann 1991, Gaumert and Kämmereit 1993, Spratte and Hartmann 1998, Winkler et al. 2002). It is known that river lampreys live predominantly in coastal areas, especially near estuaries (Hardisty 1986), while sea lampreys are also found in the open sea up to several hundred kilometres away from the coast (Lelek 1973). When mature, the lampreys migrate upstream the rivers. Spawning takes place only in freshwater, mostly far upstream. Historically, lampreys were mainly caught during their spawning migrations (Sterner 1918, Imam et al. 1958). Generally, lamprey populations have decreased in Germany since the mid-1950s (e.g., Imam et al. 1958, Wilkens and Köhler 1977, Möller 1984). It is assumed that pollution of estuaries and the building of weirs and dams prevented lampreys from reaching their original spawning grounds (Lelek 1987).

Historically, allis shad populations occurred along the eastern Atlantic coasts from Norway to Morocco and into the western Mediterranean Sea, extending along the coasts of Portugal, Spain, France, British Isles, Belgium, Holland and Germany (Baglinière et al. 2003). In Germany, allis shad was an economically important species, especially in the basin of the River Rhine, prior to the beginning of the 20th century. Fishing has been invoked as one of the primary factors involved in the reduction of the Rhine population of allis shad (de Groot 1989). According to Bartl and Troschel (1997), massive overfishing during the beginning and heavy pollution in the middle of the 20th century may be the reasons why allis shad vanished from the River Rhine. Although the last specimen from the River Rhine was caught in 1963 (Bartl and Troschel 1997), the species disappeared even earlier from other German rivers draining into the North Sea and Baltic Sea basins (e.g., Duncker 1935–1939, Freyhof 2002) due to poor water quality and building of weirs (Lelek 1987). The number of recorded allis shad, however, has increased in the River Rhine since 1978 (Grimm 1993, Freyhof 2002).

In contrast, the subspecies *A. fallax fallax* of twaite shad is distributed in the Baltic Sea and along the entire Atlantic sea coast, including the North Sea (e.g., Saemundsson 1949, Kartas 1981, Taverny 1991, Sabatié 1993, Aprahamian et al. 2003). However, the correct nomenclature for the twaite shad population(s) in the Atlantic and the Baltic Sea has not yet been finally determined. Winkler et al. (2000) evaluated the distribution status of twaite shad within the framework of their checklist of fish species in the Baltic Sea. The authors described the distribution status of *A. fallax fallax* in the Baltic Sea as "present" for Denmark, "common" for Poland and Lithuania, and as "very rare" for Russia, Finland, Latvia and Germany. Generally, the twaite shad, which was very common in a number of Baltic and other European waters about a hundred years ago, has declined substantially throughout Europe (Reshetnikov et al. 1997). This decline has been attributed to pollution, overfishing, and migratory route obstructions (Whitehead 1985). Around 1990 there were only a few rivers left with healthy populations of twaite shad, like the Garonne–Dordogne river system in France and the Elbe River (North Sea) in Germany (Quignard and Douchment 1991).

Due to their limited salinity tolerance, the occurrence of freshwater Annex II-fish species is restricted to freshwater and oligohaline-mesohaline habitats of the German North Sea and Baltic Sea regions (e.g.,

Thiel et al. 1995, Thiel and Potter 2001, Thiel 2003). Details of their status will not be discussed here. General information about their distribution and habitat requirements in the German North- and Baltic Sea basins are presented in Diercking and Wehrmann (1991), Gaumert and Kämmereit (1993), Spratte and Hartmann (1998), Fricke (2000), Freyhof (2002), and Winkler et al. (2002).

3 Distribution of Annex II-fish species in German waters of the North Sea

In May and August 2002, the BFAFi carried out a study to describe the present distribution of the ichthyofauna including Annex II-fish species in the German parts of the North Sea. During that study, 61 beam trawl hauls were performed in four study areas. Furthermore, an analysis of the historical distribution of the relevant fish species in the whole German EEZ was carried out using data from the BFAFi research fishing campaigns from 1978 to the present. During this period, 3,629 hauls were performed with different trawls, and they were analysed for the occurrence of Annex II-fish species (Kloppmann et al. 2003).

The data analysis showed no concentration areas (high abundances) for any of the Annex II-fish species (i.e., twaite shad and lamprey species) in the German EEZ. There have been no records of sturgeon, North Sea houting, allis shad or of any freshwater Annex II-fish species in the German EEZ since 1981 (Kloppmann et al. 2003). The sturgeon is considered extinct as a reproductive species in the North Sea region (Kirschbaum and Gessner 2002). The same is true for the houting; only a small population has survived in the River Vida (Denmark) where the species was rediscovered in 1982. A restitution programme, based on transferred fishes from the River Vida to the River Treene, the River Elbe basin, and to the lower River Rhine in Germany, has been running since the end of the 1980s (Jäger 1999). However, none of these stocking projects has resulted in a self-sustainable population independent of stocking. Recent findings show that the relict stocks of houting from the North Sea basin are identical to the houtings living in the southern Baltic and belong to the species *Coregonus maraena*. The original North Sea houting (*Coregonus oxyrinchus*) is a globally extinct species (Freyhof and Schöter 2005).

Today the main distribution range of allis shad is restricted to the Atlantic coasts of France and Portugal (Baglinière et al. 2003) and it is possible that those individuals found in the River Rhine are only vagrants from the large French populations (Freyhof 2002). Therefore, the species

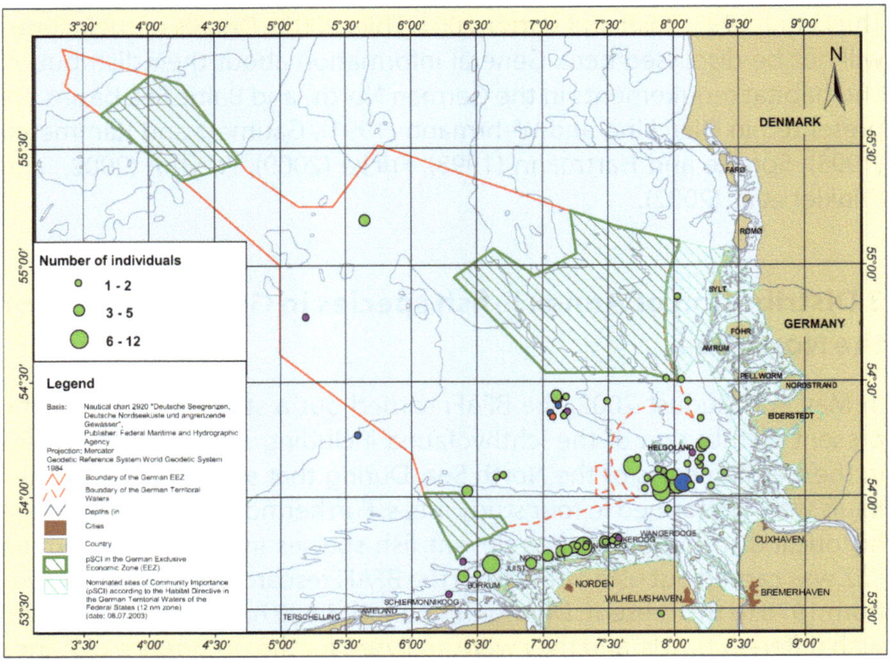

Figure 3. Distribution of lamprey catches (*L. fluviatilis* and *P. marinus*) from BFAFi in the North Sea from 1978 until 2002 according to Kloppmann et al. (2003). Purple: first quarter; red: second quarter; blue: third quarter; green: fourth quarter

is expected to occur only accidentally in the German waters of the North- and Baltic Seas.

Since 1978, 178 records of lampreys have been obtained in the German North Sea. Around 80% originated from approximately 1995 onwards (Kloppmann et al. 2003). Most of the lamprey records (mainly river lampreys) were located in nearshore areas, especially between Helgoland and the mouth of the Elbe and Weser estuaries (figure 3).

Furthermore, most of these records were obtained between July and September, which could be attributed to nearshore concentrations of river lampreys due to their spawning migration. Thiel and Salewski (2003) estimated that the anadromous spawning migration of river lampreys into the Elbe Estuary proceeds in autumn. A spring spawning migration of river lampreys was not observed. In contrast, sea lampreys migrated for spawning mainly during spring (Thiel and Salewski 2003). However, only 10 individuals of sea lampreys in the Elbe Estuary were recorded from 1989–1995 (Thiel and Salewski 2003). In comparison, 2,217 river lampreys were caught in the same area during the same period. This indicates

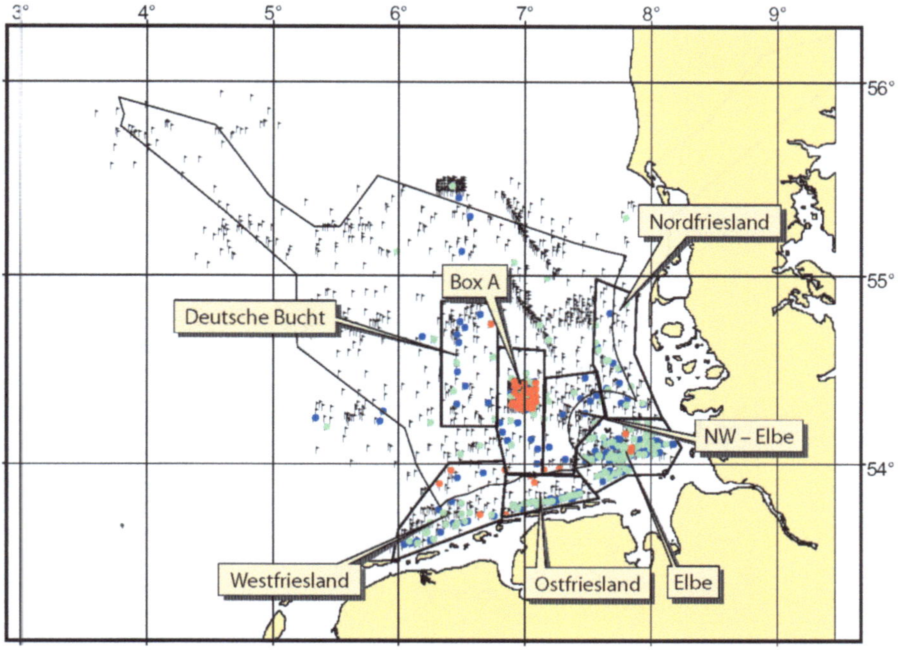

Figure 4. Distribution of all hauls and twaite shad catches from BFAFi from 1978 until 2002 according to Kloppmann et al. (2003). Flags represent individual hauls. Framed boxes: defined areas for evaluation of the spatial effort/catch proportion. Blue: first quarter; red: third quarter; green: fourth quarter. No records of twaite shads occurred in the second quarter

remarkable differences in the population densities of both lamprey species. Sea lampreys have never been very abundant in the southern North Sea region, e.g., in the Elbe Estuary (Kluge 1904, Bauch 1958).

Within the German coastal waters (12-nautical mile zone), there were high-density areas of twaite shad. In total, 6,052 individuals of twaite shad have been caught in the *German Bight* since 1978 (figure 4). Figure 4 demonstrates that apart from the 12-nautical mile zone, there were also several twaite shad records in the German EEZ. Since the spatial effort/catch proportion differs between the framed areas, then their records of twaite shad are not alike or comparable. For example, the effort within the areas "Box A" and "Elbe" was higher than in the remaining areas (Kloppmann et al. 2003).

Stelzenmüller and Zauke (2003) focused their study on the distribution of twaite shad. They analysed catch data from 1986–2001 provided by the BFAFi. The data analysis was performed with geo-statistical methods

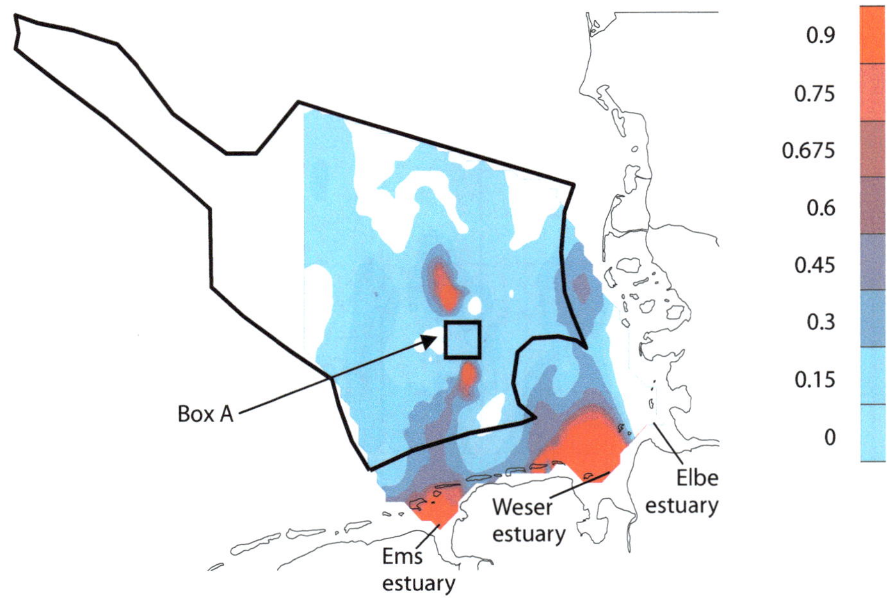

Figure 5. Above: Calculated probability (indicator kriging) to catch at least one twaite shad within the EEZ and coastal waters between 1987 and 2001 according to Stelzenmüller and Zauke (2003), identifying proven concentration areas in the Ems and Weser estuaries. Below: relative measure of uncertainty of the estimate using indicator kriging

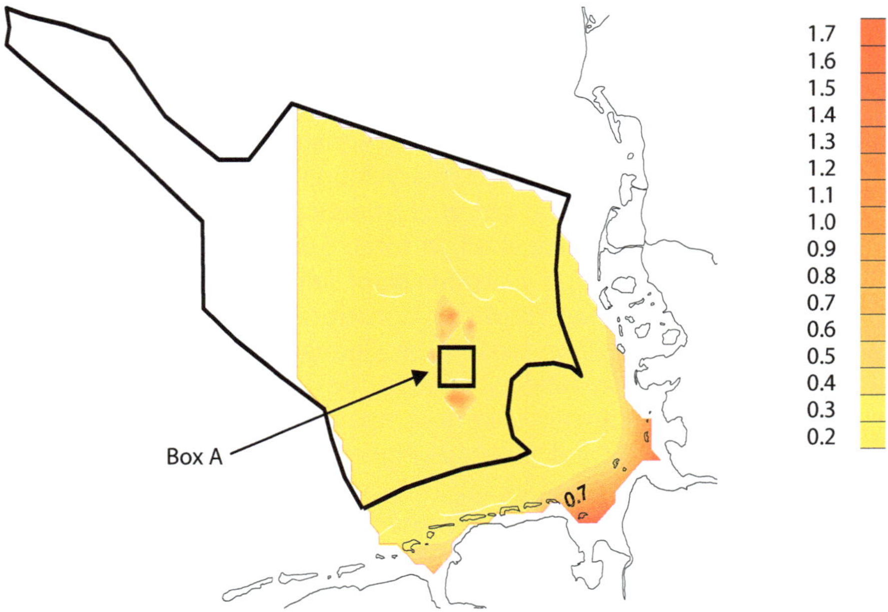

and modelling programs in order to identify possible concentrations of twaite shad within the German North Sea. The data evaluation showed that in 166 out of 1,310 sampled stations (12.67%) at least one individual twaite shad was caught. Since 1995, however, the catch effort and the percentage of stations with twaite shad records increased and in general, there were more records of this species in the German coastal waters than in the German EEZ. The analysis indicates that the population of twaite shad has increased in the study area since 1997. The spatial modelling of the distribution patterns of twaite shad and the smoothing of the patterns using indicator kriging[1] indicate the importance of the coastal waters for this species (figure 5). Figure 5 also illustrates the relative measure of uncertainty of the estimation using indicator kriging and shows that the higher occurrences around the area "Box A" cannot be verified (for further details see Stelzenmüller and Zauke 2003).

The increased uncertainty within the outer parts of the Weser and Elbe estuaries (see figure 5) is caused by the geometry of the investigation area; therefore, the former has no influence on the interpretation of the results (Stelzenmüller and Zauke 2003). In particular, the tidally influenced outer parts of the Ems, Weser, and Elbe estuaries show clear aggregations of twaite shads (figure 5). Within these areas, the probability of catching a specimen of twaite shad is approximately 90%, whereas there were no aggregation areas in the EEZ (Stelzenmüller and Zauke 2003). Stelzenmüller et al. (2004) confirmed the importance of the Weser and Elbe estuaries as areas with the highest probabilities of catching twaite shad, while within the German EEZ of the North Sea, no such areas could be discerned. Spawning populations of twaite shad exist in the estuaries of Elbe and Weser (e.g., Hass 1965, Möller and Dieckwisch 1991, Scheffel and Schirmer 1991, Thiel et al. 1996, Gerkens and Thiel 2001). Actual spawning activities have not yet been observed in the estuaries of Ems and Eider, although adult individuals were caught in the River Eider (Vorberg and Breckling 1999) and high abundance of small juveniles occur in the Wadden Sea close to the mouth of the Eider Estuary during summer (Breckling et al. 1994). Freshwater Annex II-fish species were not found within the German EEZ and coastal waters of the North Sea. However, asp, spined loach, bitterling and white-finned gudgeon were caught, for instance, in the Elbe Estuary (Thiel et al. 1995, Thiel and Bos 1998, Jankowski 2001, Thiel and Potter 2001, Schubert 2004). From these

[1] Kriging: a form of statistical modelling that interpolates data from a known set of sample points to a continuous surface (spatial distribution).

species, only the asp occurred frequently in oligohaline-mesohaline water bodies.

In conclusion, with the available data, no concentration areas for Annex II-fish species could be shown for the EEZ; however, for the twaite shad such concentrations are evident in the Wadden Sea, particularly in the outer areas of the Ems, Elbe and Weser estuaries. This species also occurs in the EEZ and its population seems to be increasing. Sea and river lampreys exist throughout the North Sea, but catches are too low to conduct significant geo-statistical analysis for spatial distribution patterns. It is, however, much more likely to catch a river lamprey than a sea lamprey in the North Sea.

4 Occurrence of Annex II-fish species in German waters of the Baltic Sea

In June 2002, the BFAFi performed in total 38 hauls using otter trawls (Kloppmann et al. 2003). In the Baltic Sea, due to the stony bottom, just one study area was sampled with 7 hauls adjacent to the actual area. The other 31 hauls were carried out in areas outside potential SACs. The historical dataset used for the analysis of the occurrence of Annex II-fish species was from 1991 and this was based on data resulting from 2,979 hauls taken with trawls. Some additional data originating from studies in the *Pommeranian Bay* and from hydro-acoustic surveys were available from 1978. The results of the recent field work and historical data analysis showed that there were no areas of high concentration of Annex II-fish species within the German EEZ of the Baltic Sea as well (Kloppmann et al. 2003). Within this area, 2 individuals of twaite shad were found as well as one individual each of sea lamprey and of houting (Kloppmann et al. 2003). However, it is of note that the monitoring techniques used (e.g., seasonal sampling regime, mesh size in cod end) were not fully appropriate to verify the occurrence of all life history stages of Annex II-fish species. Furthermore, in areas with rocky bottom or stony reefs, no catches were carried out due to problems with operating the trawls. The results indicate a lack of data for most of the Annex II-fish species in the German EEZ and show the urgent need to develop and employ alternative research methods in order to get more detailed data on the occurrence and distribution of Annex II-fish species within the German EEZ.

In order to remedy this gap in knowledge, in 2003 the German Oceanographic Museum (DMM) and the University of Rostock (IfB) started a study in order to test more specialised data enquiries from existing sources and to develop new sampling strategies for Annex II-fish species in the Baltic Sea, with special attention to the *Pommeranian Bay* and adjacent waters (Thiel et al. 2004a, 2004b, 2005). From August 2003 to August 2004, the first results of this study were obtained based on the analysis of relevant ichthyological collections and the compilation of recent catch records from fisheries research, and commercial and recreational fishery in the German parts of the Baltic Sea. Furthermore, catch data from research hauls in the Baltic Sea region between the Islands of Rügen, Usedom, and Bornholm were evaluated by DMM and IfB using special equipment.

In order to obtain information regarding the historical occurrence of Annex II-fish species, the ichthyological collections of 17 institutions were analysed, including those of the Museum of Natural History of the Humboldt University in Berlin, Zoological Institute and Museum of the University of Hamburg, Museum of Natural History Stuttgart, Zoological Museum of the Christian Albrechts University in Kiel, Biology Department of the University of Rostock, German Oceanographic Museum in Stralsund, and the Zoological Institute and Museum of the Ernst Moritz Arndt University Greifswald.

The analysis of these collections resulted in 1,435 spatially and temporally different records of Annex II-fish species, including *A. oxyrinchus* and *C. maraena* from 1822–1999. Apart from white-finned gudgeon, historical records of all of the relevant Annex II-fish species were identified in the Baltic Sea region (Thiel et al. 2004a, 2004b, 2005). About 32% of the recorded individuals date from the 19th century, and 65% from the 20th century. Most records obtained were for twaite shad (384 records), houting (234 records), and river lamprey (219 records). Thiel et al. (2004b) found that the oldest records date back to 1822, which were 2 Atlantic sturgeons (caught in the *Greifswald Lagoon* (*Greifswalder Bodden*) and close to Hiddensee Island), 1 bullhead from the same lagoon, and 1 weatherfish originating from waters around the city of Barth. The record of allis shad in the *Strelasund* in 1998 was the only specimen of *A. alosa* caught in German Baltic waters during the last 20 years (Winkler et al. 2002, Thiel 2003).

About 6 freshwater Annex II-fish species were recorded within the coastal waters of the German Baltic areas. These species are asp, spined loach, ziege, bullhead, bitterling and weatherfish. Most records were obtained from the *Greifswald Lagoon* and *Szczecin Lagoon* and its adjacent waters in the period from 1822 to 1999 (figure 6). Only two freshwater species, namely asp and spined loach, were estimated with high number of records, especially in the *Greifswald Lagoon* and *Szczecin Lagoon*. The spined loach was also found in the *Schlei Fjord*, River Trave system and Warnow River (figure 6). Schaarschmidt and Lemcke (2004) recorded asps also from the Darss-Zingst Estuary, whereas Winkler et al. (2002) excluded the presence of asp in that estuary. Recently, Lorenz (2001) reported asp and spined loach presence in shallow habitats of the *Szczecin Lagoon* as well. Winkler et al. (2002) also estimated the presence of spined loach in the Darss-Zingst Estuary.

Apart from freshwater habitats, the occurrence of the freshwater Annex II-fish species was mainly restricted to oligohaline-mesohaline coastal lagoons and estuaries. However, the ziege was also recorded in coastal waters east of the Rügen Island (figure 6). Nowadays, this species is known as a rare visitor to German coastal waters and lagoons east of Rügen Island (Spieß and Waterstraat 1989, Winkler 1989). Schaarschmidt and Lemcke (2004) also found historical records of this species for coastal waters west of the island.

The white-finned gudgeon has not been observed in German Baltic waters yet, although there exists a population of this species in the river system of the *Odra* (Rolik 1965, Blachuta et al. 1994, Freyhof et al. 2000). According to Naseka and Freyhof (2004), the white-finned gudgeon of the *Odra* was recently classified as *Romanogobio belingi* (Slastenenko, 1934); however, the white-finned gudgeon is still listed as *Gobio albipinnatus* (Lukasch, 1933) in the Annex II of the Habitats Directive.

From 1822–1999, the highest records of river and sea lampreys were obtained from the waters around the Rügen Island, the *Szczecin Lagoon* with adjacent waters, from the mouth of the Warnow River, the *Mecklenburg Bay* (*Mecklenburger Bucht*) and from the *Kiel Bight* (*Kieler Bucht*) (figure 7). This indicates the importance of these waters as habitat for lampreys at that time. Important lamprey stocks also existed east of these areas. From 1900–1920, more than 30 tons of river lampreys were caught annually in the *Gdańsk Bay* (*Danziger Bucht*), the *Vistula Lagoon* (*Frisches Haff*), and the *Curonian Lagoon* (*Kurisches Haff*) (Anonymus 1900–1920).

Twaite shad was also an important commercial species in some areas of the southern Baltic, especially during the last quarter of the nineteenth- and the first half of the twentieth century (Thiel et al. 2004a). The mean annual twaite shad catch from 1891 to 1960 amounted to 87 tons for the entire southern Baltic Sea. The annual catches of this species in this area declined sharply in the 1950s so that twaite shad was only occasionally caught in the Baltic Sea region until the mid-1990s (Thiel et al. 2004a).

The present occurrence of Annex II-fish species in the Baltic Sea region was investigated via a research fishery with otter trawls and shrimp trawls.

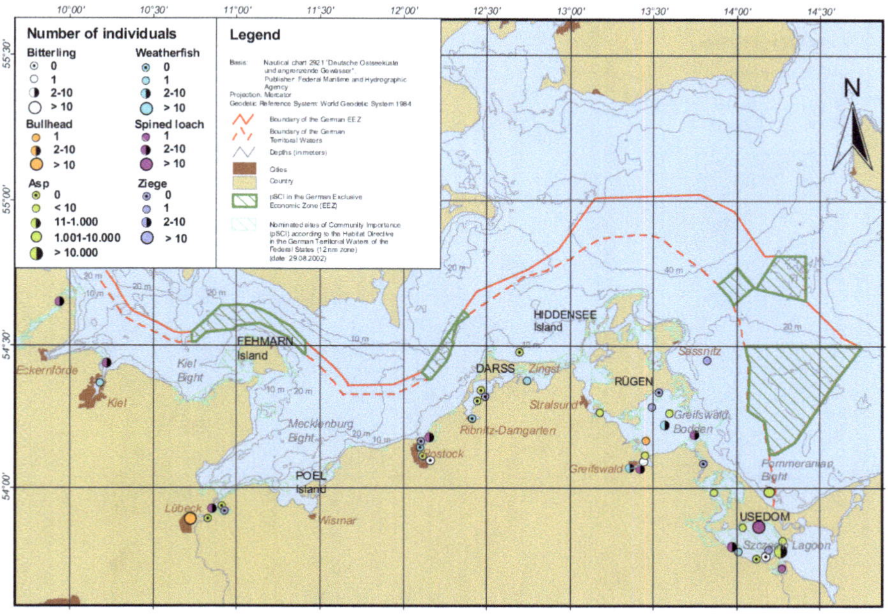

Figure 6. Distribution of historical records of freshwater Annex II-fish species in German Baltic waters from 1822–1999

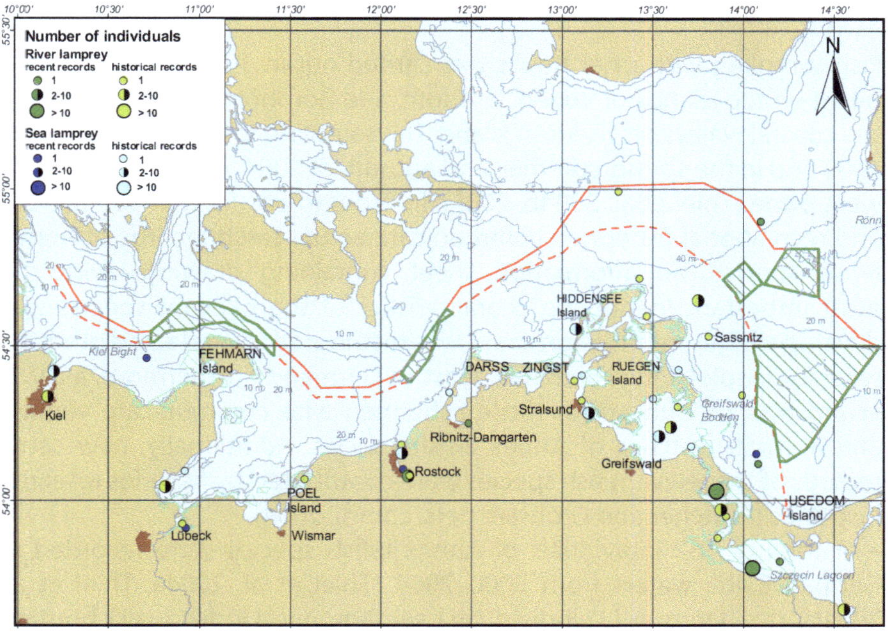

Figure 7. Distribution of recent records (since 2000) and historical records (1822–1999) of river and sea lampreys in German Baltic waters (legend in figure 6)

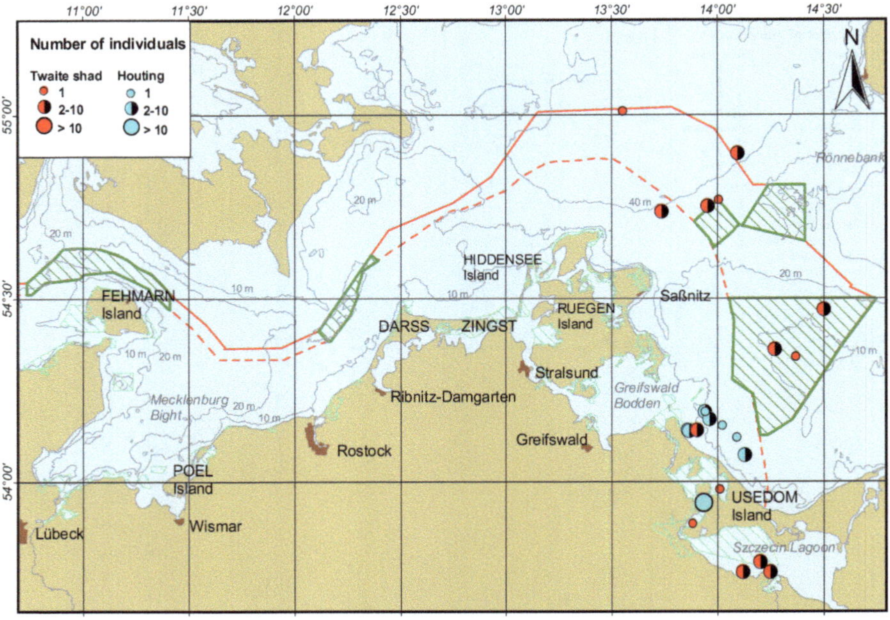

Figure 8. Distribution of recent records of twaite shad and houting in German Baltic waters from May 2003 until November 2004 (legend in figure 6)

Fish sampling with otter trawls was carried out in 3 investigation areas between the Islands of Rügen, Usedom, and Bornholm in the EEZ and in the coastal waters off Usedom Island. The sampling took place in autumn 2003 and in the spring, summer, and autumn of 2004. A total of 108 hauls were done (Thiel et al. 2004b, 2005). In order to ensure that commercial and recreational fishermen communicate actual catch records of Annex II-fish species, an information sheet containing drawings with the most important identifying characteristics of the relevant species was developed. This information sheet was distributed to a selected group of reliable people along the Baltic coast of Germany. Furthermore, a catch award was announced for those fishermen who communicate verified and accurate catches of Annex II-fish species. Additionally, new catch records of the relevant fish species were contributed by fisheries research institutes (Böttcher and Gröhsler, pers. comm. 2004).

A total of 162 individuals of Annex II-fish species were recorded in German Baltic waters from 2000–2004 (Thiel et al. 2004a, Thiel et al. 2004b): river lamprey (81 individuals), sea lamprey (4 individuals), houting (38 individuals), twaite shad (38 individuals) and *Alosa* sp. (1 individual). These records were obtained from 36 different localities, although 64%

of these spatial records originated from northeast of Rügen Island, the *Pommeranian Bay*, and the *Szczecin Lagoon* and adjacent waters (figures 7 and 8). Eighty-four percent (84%) of the river lamprey individual records originated from the *Szczecin Lagoon* and its adjacent waters (figure 7), demonstrating the importance of these bodies of water for the spawning migrations of this species. In comparison, only 4 single individuals of sea lamprey were caught at 4 different locations in German Baltic waters. Nowadays, no actual reproduction of sea lamprey in the German Baltic Sea area is known (Spratte and Hartmann 1998, Winkler et al. 2002). Historically, the sea lamprey may have spawned in the River Trave system (Duncker 1935–1939). No regular annual spawning of river lamprey occurs at all spawning sites in the German Baltic Sea region (Winkler et al. 1999). Additionally, these spawning populations are very small, comprising only of 20 to 100 individuals (Winkler et al. 2002).

Most of the spatial records of houting (80%) were obtained from the coastal waters off the Usedom Island (figure 8), indicating a nearshore distribution of this species in the *Pommeranian Bay*. Spawning concentrations occur in the *Szczecin Lagoon* and adjacent waters (figure 8). Kottelat (1997) associated the anadromous houting stocks of the German Baltic waters with *Coregonus maraena* (Bloch, 1779) from Lake Madü in Poland. According to Freyhof and Schöter (2005), the houtings from the Rivers Ems, Elbe, Treene, Schlei, Peene, and from the *Schlei Fjord* and the *Vänern* also belong to the same species. Although the population has stabilised during the last 10 years (Winkler et al. 2002), and restitution programmes have been underway, (for example, in the River Trave and in the *Schlei Fjord* using *C. maraena* from the *Szczecin Lagoon* and adjacent waters since 1992) the species is very close to disappearing from several German Baltic waters (Freyhof 2002). However, the main Baltic distribution area of houting is the *Szczecin Lagoon* and adjacent waters (Schulz 2001). A stocking programme was running there from 1996 to 2002, and it has resumed since 2005.

From May 2003 until November 2004, twaite shads were found at 14 different localities in the German EEZ northeast of the Rügen Island, as well as in the *Pommeranian Bay* and the *Szczecin Lagoon* and its adjacent waters (figure 8). With a total of 38 individuals, this species contributed 23% of all recent records of Annex II-fish species. All 19 records of twaite shad (50%) from the *Szczecin Lagoon* and adjacent waters were adult individuals, and dated from May to July (Thiel et al. 2004a, 2005). In the *Odra Bank* and in the coastal waters off the Usedom Island, only juvenile individuals of age group 0 were caught amounting to 19% of the total number of records. Thirty-one percent (31%) of all individuals of twaite shad (mainly adults) were recorded from the *Western Rønne Bank* and adjacent waters.

Given the recent records from the German Baltic waters, it is assumed that the Baltic population of twaite shad has been increasing since the middle of the 1990s, after about 50 years of decline. Migration of greater numbers of twaite shads from the North Sea into the Baltic Sea has not been observed yet. On the other hand, the species has also been observed more frequently in the Polish, Lithuanian, and Estonian waters of the southern Baltic Sea since the mid-1990s. Therefore, the source of the population increase could be the eastern twaite shad stock of the *Curonian Lagoon* (Thiel et al. 2004a).

5 Conclusions

The status and population trends of Annex II-fish species (especially the population increase of twaite shad) reported here indicate that it is important to investigate the future population dynamics of such fish species in German waters of the Baltic Sea and North Sea. In particular, the study of the distribution of twaite shad and river lamprey, the most important Annex II-fish species in the German waters, needs to be continued. An international cooperation with the new EU Member States of the southern Baltic should be initiated to estimate the overall status of the Annex II-fish species for the southern Baltic. The importance of the *Szczecin Lagoon* (Baltic Sea) and of the Ems and Eider estuaries (North Sea) as spawning and nursery habitats of twaite shad needs further investigation to allow a more robust evaluation of the EEZ and the coastal waters of Germany as habitat for this species.

New methods (e.g., underwater video techniques, SCUBA diving) should be employed to get more accurate data on the occurrence of lampreys in stony reef habitats. Clearly further investigations regarding the systematics of migratory Coregonids (houtings) in the North- and Baltic Seas are necessary to clarify the status of some of the Habitats Directive's Annex II-fish species populations.

Acknowledgements

We thank Renate Neumann, Philip Riel, Helmut Winkler, Uwe Böttcher, Bernd Bruns, Tomas Gröhsler, Jens Heischkel, Volker Huckstorf, Dietmar Lill, Nardine Löser, Ronny Weigel, and two anonymous reviewers for their helpful contributions and comments to this study. The study was assigned by the German Federal Agency for Nature Conservation and sponsored by the Federal Ministry for the Environment, Nature Conservation and Nuclear Safety under grant number 80385220.

References

Anonymous (1900–1920) Jahresberichte über die deutsche See- und Küstenfischerei. W. Moeser Hofbuchhandlung, Berlin

Aprahamian MW, Baglinière J-L, Sabatié MR, Alexandrino P, Thiel R, Aprahamian CD (2003) Biology, status, and conservation of the anadromous Atlantic twaite shad *Alosa fallax fallax*. American Fisheries Society Symposium 35:103–124

Baglinière J-L, Sabatié MR, Rochard E, Alexandrino P, Aprahamian MW (2003) The allis shad *Alosa alosa*: biology, ecology, range and status of populations. American Fisheries Society Symposium 35:85–102

Bartl G, Troschel HJ (1997) Bestandsentwicklung und aktuelle Situation von *Alosa alosa* und *Alosa fallax* im Rheingebiet. Zeitschrift für Fischkunde 4:119–162

Bauch G (1958) Untersuchungen über die Gründe des Ertragsrückganges der Elbfischerei zwischen Elbsandsteingebirge und Boizenburg. Z.f. Fischerei N.F. 7:161–437

Błachuta J, Kotusz J, Witkowski A (1994) The first record of the whitefin gudgeon, *Gobio albipinnatus* Lukasch, 1933, (Cyprinidae), from the Odra River basin. Prezeglad Zoologiczny XXXVIII:3–4

Breckling P, Beermann-Schleiff S, Achenbach I, Opitz S, Walthemath M, Nellen W, Berghahn R, Schnack D (1994) Fische und Krebse im Wattenmeer. Study for the Federal Environmental Agency. University of Hamburg, Christian-Albrechts-University Kiel. Final report 12/1994, Volume 1

Debus L (1995) Historic and recent distribution of *Acipenser sturio* in the North Sea and Baltic Sea. Proceedings of the International Sturgeon Symposium, VNIRO, Moscow, pp 189–203

Diercking R, Wehrmann L (1991) Artenschutzprogramm Fische und Rundmäuler in Hamburg. Naturschutz und Landschaftspflege in Hamburg 38:1–126

Duncker G (1935–1939) Die Fische der Nordmark, 1-5. Lieferung, Kommissions-Verlag Lipsius & Tischer, Kiel

Elliott M, Hemingway KL (2002) Fishes in Estuaries. Blackwell Publishing, Oxford

Freyhof J (2002) Freshwater fish diversity in Germany, threats and species extinction. In: Collares-Pereira MJ, Coelho MM, Cowx IG (eds) Freshwater fish conservation: options for the future. Blackwell Publishing, Oxford, pp 3–22

Freyhof J, Schöter C (2005) The houting, *Coregonus oxyrinchus* (L.) (Salmoniformes: Coregonidae), a globally extinct species from the North Sea basin. Journal of Fish Biology 67:713–729

Freyhof J, Scholten M, Bischoff A, Wanzenböck J, Staas S, Wolter C (2000) Extensions to the known range of the whitefin gudgeon *Gobio albipinnatus* Lukasch, 1933, in Europe and biogeographical implications. Journal of Fish Biology 57:1339–1342

Fricke R (2000) Auswahl und Management mariner NATURA-2000-Gebiete für Fischarten im Anhang II der FFH-Richtlinie. In: Petersen B, Hauke U, Ssymank A (eds) Der Schutz von Tier- und Pflanzenarten bei der Umsetzung der FFH-Richtlinie. Schriftenreihe für Landschaftspflege und Naturschutz 68:113ff

Gaumert D, Kämmereit M (1993) Süßwasserfische in Niedersachsen. Niedersächsisches Landesamt für Ökologie, Hildesheim

Gerkens M, Thiel R (2001) Habitat use of age-0 twaite shad (*Alosa fallax* Lacépède, 1803) in the tidal freshwater region of the Elbe River, Germany. Bulletin Français de la Pêche et de la Pisciculture 362/363:773–784

Grimm R (1993) Fische und Fischerei im Oberrhein. Berichte z. Fischereiforschung 3, Fischereiforschungsstelle d. Landes Bad.-Württ., Langenargen

de Groot SJ (1989) The former allis and twaite shad fisheries of the lower Rhine, The Netherlands. ICES C.M. 1989/M:19

Hardisty MW (1986) A general introduction to lampreys. In: Holcik J (ed) The Freshwater Fishes of Europe. Band 1/1 Petromyzontiformes. Aula Verlag, Wiesbaden

Hass H (1965) Untersuchungen über den Laichfischbestand der Elbfinte, *Alosa fallax* (Lacépède 1803). Archiv für Fischereiwissenschaft 16:150–168

Imam AKES, Lühmann M, Mann H (1958) Über Neunaugen und Neunaugenfischerei in der Elbe. Fischwirt 9:249–261

Jäger T (1999) Die Wiedereinbürgerung des Nordseeschnäpels. In: Verband Deutscher Sportfischer (ed) Der Nordseeschnäpel. Selbstverlag Verband Deutscher Sportfischer, Offenbach, pp 3–11

Jankowski R (2001) Nahrung und Habitatwahl von Rapfen *Aspius aspius* (Linaeus, 1758) der Altersgruppe 0 in der limnischen Tideelbe. MSc thesis, University of Hamburg

Kartas F (1981) Les clupéidés de Tunisie. Caractéristiques biométriques et biologiques. Etude comparée des populations de l'Atlantique est de la Méditerranée. In: The Clupeidae of Tunisia: Their biometrical and biological characteristics. A comparative study of eastern Atlantic and Mediterranean populations. Thèse de Doctorat d'Etat. Tus, Tunisia: Université de Tunis

Kirschbaum F, Gessner J (2002) Perspektiven der Wiedereinbürgerung des Europäischen Störs, *Acipenser sturio* L., im Einzugsgebiet der Elbe. Zeitschrift für Fischkunde, Suppl. 1:217–232

Kloppmann MHF, Böttcher U, Damm U, Ehrich S, Mieske B, Schultz N, Zumholz K (2003) Erfassung von FFH-Anhang II-Fischarten in der deutschen AWZ von Nord- und Ostsee. Study for the German Federal Agency of Nature Conservation. Federal Research Centre for Fisheries, Hamburg. Final report 11/2003

Kluge M (1904) Zum Neunaugenfang am Cracauer Wehr bei Magdeburg. Fischerei-Zeitung Neudamm 7:485

Kottelat M (1997) European Freshwater Fishes. An heuristic checklist of the freshwater fishes of Europe (exclusive of former USSR), with an introduction for non-systematics and comments on nomenclature and conservation. Biologia 52 Suppl. 5:1–271

Lelek A (1973) Occurrence of the Sea Lamprey in midwater of Europe. Copeia 1:136–137

Lelek A (1987) Threatened Fishes of Europe. In: Holcik J (ed) The Freshwater Fishes of Europe. Volume 9, Aula Verlag, Wiesbaden

Lorenz T (2001) Aufkommen, Verteilung und Wachstum von Fischlarven und Jungfischen im Kleinen Stettiner Haff unter besonderer Berücksichtigung der wirtschaftlich wichtigen Arten Zander (*Stizostedion lucioperca* (L.)) und Flußbarsch (*Perca fluviatilis*

L.). Fisch und Umwelt Mecklenburg-Vorpommern e.V., Jahresheft 2001:21–45
Ludwig A, Debus L, Lieckfeldt D, Wirgin I, Benecke N, Jenneckens I, Williot P, Waldman JR, Pitra C (2002) When the American sea sturgeon swam east. Nature 419:447–448
Möller H (1984) Daten zur Biologie der Elbfische. Verlag Heino Möller, Kiel.
Möller H, Dieckwisch B (1991) Larval fish production in the tidal River Elbe 1985–1986. Journal of Fish Biology 38:829–838
Naseka AM, Freyhof J (2004): *Romanogobio parvus*, a new gudgeon from River Kuban, southern Russia (Cyprinidae, Gobioninae). Ichthyol. Explor. Freshw. 15:17–23
Quignard JP, Douchement C (1991) *Alosa fallax fallax* (Lacépède, 1803). In: Hoestlandt H (ed) The freshwater fishes of Europe. Volume 2. Clupeidae, Anguillidae 2. Aula-Verlag, Wiesbaden, pp 225–253.
Reshetnikov YS, Bogutskaya NG, Vasil'eva ED, Dorofeeva EA, Naseka AM, Popova OA, Savvaitova KA, Sideleva VG, Sokolov LI (1997) An annotated check-list of the freshwater fishes of Russia. J Ichthyol 37:687–736
Rolik H (1965) *Gobio albipinnatus* Lukasch. A new species to the polish fauna. Fragmenta faunistica 12 (12):177–181
Sabatié MR(1993) Recherches sur l'écologie et la biologie des aloses du Maroc (*Alosa alosa* Linné, 1758 et *Alosa fallax* Lacépède, 1803). Exploitation et taxinomie des populations atlantiques; bioécologie des aloses de l'oued Sebou. Thèse de Doctorat, U.B.O., Brest
Saemundsson B (1949) Zoology of Iceland. IV. Marine Pisces
Schaarschmidt T, Lemcke R (2004) Quellendarstellungen zur historischen Verbreitung von Fischen und Rundmäulern in Binnengewässern des heutigen Mecklenburg-Vorpommerns. Mitteilungen der Landesforschungsanstalt für Landwirtschaft und Fischerei Mecklenburg-Vorpommern 32:1–261
Scheffel H-J, Schirmer M (1991) Larvae and juveniles of freshwater and euryhaline fishes in the tidal River Weser at Bremen, FRG. Verhandlungen der Internationalen Vereinigung für Theoretische und Angewandte Limnologie 24:2446–2450
Schubert H-J (2004) Kontrolluntersuchungen im Fischaufstieg am Elbewehr bei Geesthacht. Arbeitsgemeinschaft zur Reinhaltung der Elbe, Hamburg
Schulz N (2001) Zur Bestandssituation des Ostseeschnäpels (*Coregonus lavaretus balticus*) im Peenestrom/Achterwasser Gebiet und im Stettiner Haff. Fisch und Umwelt Mecklenburg-Vorpommern e.V., Jahresheft 2001:46–63
Spieß HJ, Waterstraat A (1989) Ergebnisse der Kartierung der Rundmäuler und Fische der DDR (einheimische Arten des Süßwassers). II. Zentrale Tagung des zentralen Arbeitskreises Ichthyofaunistik der Gesellschaft für Natur und Umwelt, Feldberg, pp 11–31
Spratte S, Hartmann U (1998) Süßwasserfische und Neunaugen in Schleswig-Holstein. Ministerium für ländliche Räume, Landwirtschaft, Ernährung und Tourismus des Landes Schleswig-Holstein, Kiel
Stelzenmüller V, Zauke G-P (2003) Analyse der Verteilungsmuster der anadromen Wanderfischart Finte (*Alosa fallax*) in der Nordsee. Study for the German Federal Agency of Nature Conservation. Carl von Ossietzky University Oldenburg. Final report 03/2003
Stelzenmüller V, Maynou F, Ehrich S, Zauke G-P (2004) Application of non-linear geostatistics as a tool to search for Special Areas of Conservation for the twaite shad, *Alosa fallax* (Lacepède, 1803) in the Southern North Sea. International Review of Hydrobiology

89:337–351

Sterner E (1918) Die niederelbische Küstenfischerei, II. Die Fischereiarten. 10. Die übrigen Fischereiarten im oberen Teil. Fischerbote 10:245–253

Taverny C (1991) Pêche, biologie, écologie des aloses dans le système Grionde-Garonne-Dordogne. Contribution à la connaissance de la dynamique des populations d'aloses (*Alosa alosa* et *Alosa fallax*) dans le système fluvio-estuarien de la Gironde. Etudes CEMAGREF, Séries Réssources en eau 4

Thiel R (2003) Ästuare – wichtige Lebensräume für Fische der Nord- und Ostsee. Meer und Museum 17:36–44

Thiel R, Bos A (1998) Fischereibiologische Untersuchung verschiedener Schlenzen an der Stromelbe in Hamburg. Study for the Environmental Agency Hamburg, Final report

Thiel R, Potter IC (2001) The ichthyofaunal composition in the Elbe Estuary: an analysis in space and time. Marine Biology 138 (3):603–616

Thiel R, Salewski V (2003) Verteilung und Wanderung von Neunaugen im Elbeästuar (Deutschland). Limnologica 33:214–226

Thiel R, Sepúlveda A, Kafemann R, Nellen W (1995) Environmental factors as forces structuring the fish community of the Elbe Estuary. Journal of Fish Biology 46:47–69

Thiel R, Sepúlveda A, Oesmann S (1996) Occurrence and distribution of twaite shad (*Alosa fallax* Lacépède) in the lower Elbe River, Germany. In: Kirchhofer A, Hefti D (eds) Conservation of Endangered Freshwater Fish in Europe. Birkhäuser Verlag, Basel, Boston, Berlin, pp 157–170

Thiel R, Riel P, Neumann R, Winkler HM (2004a) Status of the anadromous twaite shad *Alosa fallax* (Lacépède, 1803) in German and adjacent waters of the Baltic Sea. ICES C.M. 2004/S:08

Thiel R, Winkler HM, Neumann R (2004b) Erfassung von FFH-Anhang II Fischarten in der deutschen AWZ von Nord- und Ostsee. Study for the German Federal Agency of Nature Conservation. German Oceanographic Museum, Stralsund. 1. Report 02/2004

Thiel R, Winkler HM, Neumann R (2005) Erfassung von FFH-Anhang II Fischarten in der deutschen AWZ von Nord- und Ostsee. Study for the German Federal Agency of Nature Conservation. German Oceanographic Museum, Stralsund. 2. Report 03/2005

Vorberg R, Breckling P (1999) Atlas der Fische im Schleswig-Holsteinischen Wattenmeer. Schriftenreihe des Nationalparks Schleswig-Holsteinisches Wattenmeer 10

Whitehead PJP (1985) Clupeoid fishes of the world – Chirocentridae, Clupeidae and Pristigasteridae. FAO Fisheries Synopsis No. 125, Volume 7, Part 1, Food and Agriculture Organization of the United Nations, Rome

Wilkens H, Köhler A (1977) Die Fischfauna der mittleren und unteren Elbe: die genutzten Arten Abh Naturwiss Ver Hamburg 20:185–222

Winkler HM (1989) Zur Fischfauna der brackigen Küstengewässer der DDR. Kenntnisstand und Gefährdung. II. Zentrale Tagung des zentralen Arbeitskreises Ichthyofaunistik der Gesellschaft für Natur und Umwelt, Feldberg, pp 32–41

Winkler HM, Spieß H-J, Waterstraat A, Krappe M, Lemcke R (1999) Monitoring von FFH-Arten von Rundmäulern und Fischen in Referenzgebieten. Naturschutzarbeit in Mecklenburg-Vorpommern 42 (1):24–40

Winkler HM, Skóra K, Repečka R, Ploks M, Neelov A, Urho L, Gushin A, Jespersen H (2000) Checklist and status of fish species in the Baltic Sea. ICES C.M. 2000/Mini 11

Winkler HM, Waterstraat A, Hamann N (2002) Rote Liste der Rundmäuler, Süßwasser- und Wanderfische Mecklenburg-Vorpommerns, kommentiert, Stand 2002. Umweltministerium Mecklenburg-Vorpommern

Chapter 10

Utilisation of time and space by harbour seals (*Phoca vitulina vitulina*) determined by new remote-sensing methods

Nikolai Liebsch[a], Rory Paul Wilson[b] and Dieter Adelung[a]

[a] *Leibniz – Institute of Marine Sciences (IFM-Geomar), Kiel*
[b] *University of Wales Swansea, Biological Sciences, Wales, UK*

Abstract

Seals in the Wadden Sea are easy to study and count when they are exposed at haulout sites. However, our knowledge of seal activity at sea is limited. Consequently, in spring and autumn of 2002 and 2003, we equipped 19 seals with a logger/PTT-combined system to record at 15-second intervals, swim speed and direction, dive depths, water temperature and water turbidity. After a predefined time, the devices were automatically released to be washed ashore, where they could be located by satellite signal or found by beach walkers. The stored data showed that the seals did not forage in the Wadden Sea, but travelled to specific *hot spots* in the North Sea where they foraged on benthic prey[1], usually for several days, before returning straight back to their sandbank. Animals almost always dived to the seabed during both commuting and foraging. However, the dive profile was more irregular in the Wadden Sea compared to the deeper North Sea where the diving pattern was very regular, particularly with respect to depth and duration. Seals in the Wadden Sea not only rested on land at their haulout spots but also *slept* in the water, sinking down to the seafloor where they lay motionless for around 7 minutes, surfacing only for a short period to breathe before sinking back again. In deeper water (over 10 metres deep), the seals rested for up to 50 minutes at the surface, apparently drifting and showing no diving activity.

[1] Benthic prey: prey living in/on the seafloor.

1 Introduction

Harbour seals (*Phoca vitulina vitulina*), together with harbour porpoises (*Phocaena phocaena*), are the most numerous marine mammals along the coast of the North Sea and they may play an important role as top predators in this ecosystem.

In the 20th century, the seal population dropped continuously due to increasing disturbance, land reclamation, pollution and particularly overexploitation, until 1976 when the hunting of seals in the whole of the Wadden Sea area was prohibited. Supported by various governmental conservation schemes, the population size of harbour seals in that area rose steadily from about 3,800 animals in1976 to approximately 10,000 animals by 1988 (Reijnders et al. 1997). However, due to an epidemic caused by the *Phocine Distemper Virus*, almost 60% of the population died at this time. Subsequently, the surviving animals gained temporary immunity[2] and the population rose again at a yearly rate of 13.3% until 2002 when about 21,000 animals were counted. Since not all animals are hauled out on sandbanks or beaches during aerial surveys, a further 30% has to be added to that number to account for the animals at sea (Schwarz 1997). The total population size at that time was thus estimated to be over 25,000 animals.

In the same year, for unknown reasons, there was another outbreak of the distemper virus in a manner similar to that in 1988. This time the population size was reduced to about 12,000 counted seals in 2003. However, already by 2003 the abundance of harbour seals had begun to rise again. We expect a fast recovery to be made similar to that experienced after 1988.

The seal expert P. Reijnders from the Netherlands analysed historic data and estimated a population size of about 37,000 animals at the beginning of the 19th century (Reijnders 1992). Under present-day conditions, taking into account intense fishery, heavy boat traffic, and other factors such as the influence of tourism, this number is unlikely to be reached again. A further new impact can arise through offshore wind farms producing disturbing noises when they are built on or nearby the feeding grounds of the seals. The impacts of offshore wind farms on seals, harbour porpoises, and birds are under investigation by the research project MINOS, funded by the Federal Ministry for the Environment, Nature

[2] Regular tests of the antibody against the PDV-Virus by the Research and Technology Centre (FTZ) of the University of Kiel (Büsum, Germany) show that the immunity of the seals lasts only 3 to 4 years. (Personal communication by Dr. U. Siebert in 2005, FTZ)

Conservation and Nuclear Safety. Some of the preliminary results of this project are presented in this paper. Given undisturbed development, a population size of around 30,000 animals in the whole Wadden Sea area seems realistic. This could be calculated from the yearly counts, published by the Common Wadden Sea Sekretariat, Wilhelmshaven.

2 Materials and methods

To be able to characterise potential dangers to the seals, such as commercial fishing, pollution, the construction and maintenance of offshore wind farms and tourism, their needs and habitats have to be known better than they are at present. Direct observations of those animals are only possible when they are on land. In the water, where they spend most of their time, researchers depend on telemetric methods to record behavioural information.

For this purpose, different techniques have been developed. VHF- and satellite-telemetry, which are excellent in terrestrial studies, have a major disadvantage in an aquatic surrounding. They only transmit information when the antenna is out of the water. Therefore, only some information could be gained with the use of these technologies in seal studies. In contrast, the satellite-compensated *dead-reckoning system* used here provides continuous recordings of all important activities of the harbour seals over weeks and months, on land and in the water. This system has been developed together with the company Driesen & Kern GmbH in Bad Bramstedt (Germany). It records information from 12 different sensors at between 3- and 20-second intervals. The memory capacity provides enough space for data to be collected over a 2-month period if parameters are sampled at, for example, 20-second intervals.

This *dead-reckoning system* stores the information from the following sensors:

- channels 1–3: compass (3D)
- channel 4: pressure (depth)
- channel 5: pressure (speed)
- channels 6–7: light intensity (2 different wavelengths)
- channels 8–9: temperature (internal/external)
- channel 10: body orientation
- channel 11: pitch
- channel 12: roll.

Figure 1. Harbour seal equipped with telemetry unit

A primary drawback of this method for the seals is that the devices have to be retrieved to access the stored data. The equipped harbour seals cannot be caught a second time to take off the unit, so the devices are incorporated in a pressure-resistant positively buoyant body which releases itself automatically after a prescribed time from a base glued to the animal (figure 1). The device is then washed ashore. The base is made of neoprene and is stuck onto the fur of the animals with a fast-setting epoxy glue. This comes off during the annual moulting of the seals.

To monitor the movement of the equipped seals and to locate the units after their release, an ARGOS satellite tag (PTT) is also part of every unit. Due to the currents in the North Sea, which run parallel to the coast, the likelihood of the floating units being washed ashore somewhere is very high. Once they are washed on land, they can be located or found by beach walkers. The recovery rate over the last 2 years was 68%. After retrieval, the stored data can be downloaded and analysed with special software. After replacing the batteries, the devices can be used again.

3 Results

During the MINOS project (TP6), 19 harbour seals were equipped and, thus far, 13 units have been successfully recovered. The data sets stored in the devices recorded between 2 to 58 days of information about the animals' behaviour. Seventeen of the animals were caught and equipped on a sandbank northwest of the Eiderstedt peninsula, the *Lorenz Plate* (*Lorenzenplate*), in the spring and autumn of 2002/2003. In addition, in December 2003, 2 animals were tagged on the Danish island of Rømø, where one unit has so far been recovered.

We have analysed the diving behaviour of more than 159 days at sea from these 13 animals, and 25 routes from foraging trips made by 6 seals. Despite the relatively small number of equipped animals, the results show a very similar pattern of animal behaviour during these spring and autumn conditions.

Diving behaviour: our harbour seals left the Wadden Sea for foraging trips in the deeper North Sea, although the animals tended to spend more time in the shallow Wadden Sea in spring. They then headed straight for the deeper waters of the open North Sea during autumn and winter.

During these trips, they dived almost continuously down to the seafloor with their dive profiles mirroring the bottom topography along their swimming route (figure 2). A closer look at the dives of harbour seals indicates that it is possible to distinguish between different activities such as swimming and foraging in shallow or deep water and resting periods.

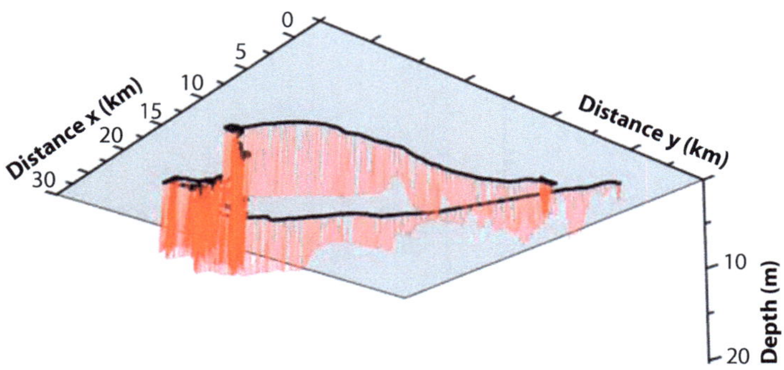

Figure 2. Route and dive profile of an equipped seal. The track to and from the foraging grounds is relatively straight, with the seal spending several days in the same area at sea.

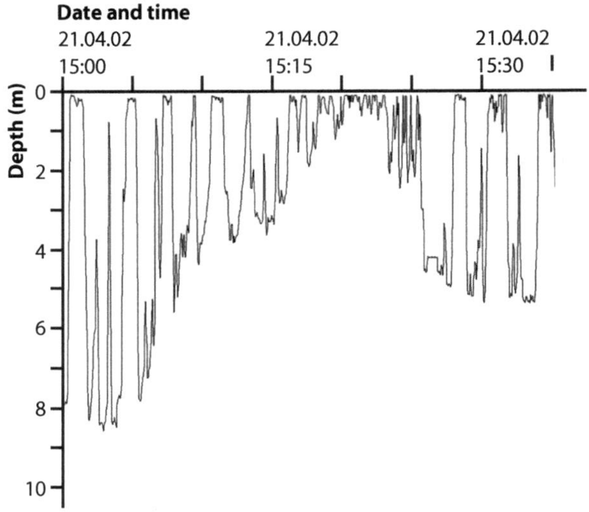

Figure 3. Irregular dive patterns in the Wadden Sea

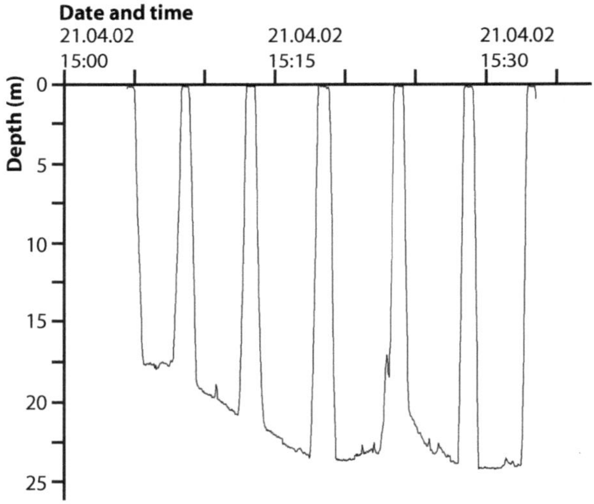

Figure 4. Dive profile at medium water depth

Swimming in the shallow water of the Wadden Sea is characterised by irregular dive patterns, independent of depth and with varying duration (figure 3).

However, foraging dives down to 20 metres, which occur mainly during the transit to deeper water, are very regular and show a clear U-shape, consisting of a vertical descent, a horizontal bottom and a vertical ascent phase. In addition, the dive profile is characterised by descent and ascent phases at a steep angle, while dive duration is around 4 minutes (figure 4).

Foraging dives become even clearer in deeper waters. Their duration is around 5 minutes, with an extended bottom phase and fast descent and ascent phases.

Apart from these 3 U-shaped dive types, other so-called *pelagic* dives occur. These have a V-shape and rarely go down to the seafloor. Animals may be searching for food during these dives.

The data also show resting periods both on land and at sea. On land, only the tilt sensors respond which show that the animals roll from one side to the other every now and then.

Astonishingly, extended resting or sleeping periods could also be found in water. In shallow areas, the seals let themselves sink down to the bottom and remain motionless for around 7 minutes, surfacing only for a short period to breathe before sinking back again.

The animals also rest from time to time while in deep water. These phases can take up to 50 minutes during which they drift to the surface without any diving activity. Such resting seals can easily be spotted from a plane during aerial surveys. Seals, diving normally, spend 79% of the time submerged at sea.

None of the presented activities was apparently influenced by the time of day or by the tides during the study period. Haul events were, however, only possible during low tide. In summer, during lactation and moulting, when no seals have ever been equipped, a higher frequency of hauling out on the sandbanks had been observed during low tide (Schwarz 1997).

Foraging trips: by using the compass, speed, depth and tilt data recorded by the device, the detailed swimming route of an equipped seal can be recalculated. This is done by starting from transmitted positions of the ARGOS-tag – while the seal is on land, before it goes into the water – and by using vectors (so-called *dead-reckoning*) for the period in water, until it comes back on land (indicated by good quality satellite fixes).

From the recorded dive data and the calculated positions, it could be presumed that the animals mainly search for food in deeper water.

One animal, which was equipped in April 2002 at the *Lorenz Plate*, swam almost directly out to the island of Helgoland the very next day. It took the seal about 16 hours to cover the distance of 57 km, while it was continuously descending, swimming along the bottom and ascending again.

Helgoland seems to be not only a permanent site for a larger population of seals in German waters, but also a temporary haulout site for seals from the mainland. In addition to the case cited above, we also detected (via satellite telemetry) another transfer to Helgoland by an adult harbour seal. Two pups equipped with ARGOS-tags upon their release from the Seal

Figure 5. Density distribution of transmitted satellite positions at the coastline of Schleswig-Holstein and the southern part of Jütland (Denmark) (The darker the colour the higher the density)

Figure 6. Density distribution of calculated route positions at the coastline of Schleswig-Holstein and the southern part of Jütland (Denmark) (The darker the colour the higher the density)

Centre in Friedrichskoog where they were raised, might also have swum to Helgoland (due to poor position quality this cannot be guaranteed) after spending considerable time on the west coast of Schleswig-Holstein.

The tagged seals primarily spent most of their time foraging in a relatively distinct area about 30 km west of the *Lorenz Plate*. It is important to mention here that quite different views on the seals' use of an area are obtained depending on whether the density calculations are based on the satellite positions (figure 5) or whether routes are calculated by dead-reckoning (figure 6). For technical reasons, only few reliable satellite data are transmitted while the seals are at sea, whereas during the haulout phases on the sandbank transmissions are not influenced. Therefore, the main seal distribution appears to be limited to that site if only the satellite telemetry data are considered. This is partly due to the positioning of the device on the back of the animal. In contrast, the continuously recorded data of the logger show that the harbour seals spend most of their time actually at sea.

Considering all analysed data, it becomes obvious that the Wadden Sea is of minor interest for hungry seals. They prefer certain *hot spots* in the

North Sea, sometimes spending days foraging out at sea, only interrupted by short resting periods.

Therefore, the seals sighted during the aerial surveys in the North Sea, which are part of the MINOS project dealing with possible effects of offshore wind farms (TP2, see chapter 14), are presumably primarily seals resting on the surface.

In the context of the offshore wind farms, harbour seals have to be given careful consideration because they obviously spend the major part of their lives foraging in the North Sea and not in the Wadden Sea.

4 Perspectives

The relevance of the results mentioned above is limited to spring and autumn. It is now important to find out if the seals' behaviour changes in the summer, during mating and moulting, and during the rest of the winter.

In addition, seal populations such as those on Helgoland and Rømø have to be examined for similar or different behaviours because they are very much influenced by their environment (Lesage et al. 1999, Fedak and Thompson 1993, Tollit et al. 1998). It would be unwise and premature to assume that the behaviour described here for seals from the *Lorenz Plate* typifies that of other populations.

References

Bowen WD, Boness DJ, Iverson SJ (1999) Diving behaviour of lactating harbour seals and their pups during maternal foraging trips. Can J Zool 77:978–988

Coltman DW, Bowen WD, Boness DJ, Iverson SJ (1997) Balancing foraging and reproduction in the male harbour seal, an aquatically mating pinniped. Anim Behav 54:663–678.

Fedak MA, Thompson D (1993) Behavioural and physiological options in diving seals. In: Boyd IL (ed) Marine mammals: advances in behavioural and population biology. Clarendon Press, Oxford, pp 333–348

Lesage V, Hammill MO, Kovacs KM (1999) Functional classification of harbor seal (*Phoca vitulina*) dives using depth profiles, swimming velocity, and an index of foraging success. Can J Zool 77:74–87

Reijnders PJH (1992) Retrospective population analysis and related future management perspectives for the harbour seal Phoca vitulina in the Wadden Sea. Neth Inst Sea Research, Publ Series 20:193–197

Reijnders PJH, Verriopoulos G, Brasseur SMJM (eds) (1997) Status of Pinnipeds relevant to the European Union. DLO Institute for Forestry and Nature Research (IBN-DLO), Wageningen

Schwarz J (1997) Untersuchungen zum Aktivitätsmuster der Seehunde (*Phoca vitulina, L.*) im schleswig-holsteinischen Wattenmeer. Entwicklung eines Korrekturfaktors für die Zählflugergebnisse. Dissertation, Universität Kiel

Tollit DJ, Black AD, Thompson PM, Mackay A, Corpe HM, Wilson B, Van Parijs SM, Grellier K, Parlene S (1998) Variations in harbour seal *Phoca vitulina* diet and dive-depths in relation to foraging habitat. J Zool London 244:209–222

Chapter 11

Evaluating the distribution and density of harbour porpoises (*Phocoena phocoena*) in selected areas in German waters

Meike Scheidat, Anita Gilles and Ursula Siebert

Research and Technology Centre Westcoast (FTZ), Büsum

Abstract

The harbour porpoise (*Phocoena phocoena*) is a small cetacean species occurring both in the German North Sea and Baltic Sea. In the process of designating marine protected areas in the framework of the European Habitats Directive (NATURA 2000), the German Federal Agency of Nature Conservation (BfN) identified candidate areas to be eventually proposed as Sites of Community Importance (pSCI). To evaluate the importance of these sites for harbour porpoises, their distribution and density were studied by conducting aerial surveys in the sites from May 2002 to September 2003 (further surveys are ongoing). Densities in the study areas were compared between study years as well as between the selected areas. The relative importance of sites was assessed by taking into account the overall distribution of porpoises in German waters. Surveys followed the standard line-transect methodology for aerial surveys. Only summer flights in the period from May to August were used for further analysis since the coverage by flights in autumn and winter was very low due to unfavourable weather conditions. In the German North Sea, 338 sightings of porpoise groups (440 individuals in total) were recorded in the summer of 2002, and 656 sightings (812 individuals in total) in the summer of 2003. In the Baltic Sea, sighting numbers in the same period were much smaller: 50 sightings (110 individuals) in 2002 and 34 sightings (43 individuals) in 2003. The main results showed clear aggregations and high densities of porpoises in the areas off the North Friesian islands of Sylt and Amrum, where there are high concentrations of the species in the summer months, which is their reproduction period. There seems to be a sharp gradient of density running from north to south. The highest density in both years was found in the study area *Sylt Outer Reef* (*Sylter Außenriff*), followed by the *Doggerbank*. Lowest densities were calculated for *Borkum Reef Ground* (*Borkum-Riffgrund*). The mean density did not differ significantly between study years in the same area. Harbour porpoise distribution in the Baltic Sea showed higher densities in the western part, namely in the *Kiel Bight* (*Kieler Bucht*) and *Flensburg Fjord* (*Flensburger Förde*), and in the eastern part close to the border of Poland. But all sightings east of the island of Rügen (study area *Pommeranian Bay* (*Pommersche Bucht*)) were only made in 2002. Thus, there is an enormous variation in the presence of habour porpoise in this area between the years. Currently surveys continue to determine how this area is used by harbour porpoises. Besides this, a clear west–east gradient in harbour porpoise density could be ascertained. The other two Baltic Sea study areas *Fehmarn Belt* (*Fehmarnbelt*) and *Kadet Trench* (*Kadetrinne*) are also used by porpoises, especially the area around the island of Fehmarn, but due to the small sizes of the areas additional investigation methods are applied, such as stationary acoustics (see chapter 12).

1 Introduction

The harbour or common porpoise (*Phocoena phocoena*) is the smallest cetacean species inhabiting temperate and cold waters throughout the northern hemisphere. Due to its occurrence, mainly in coastal or shelf waters, the porpoise is threatened by a variety of anthropogenic impacts including by-catch in fishery (Vinther 1999, ASCOBANS 2000) and habitat degradation due to, for example, chemical pollution (Jepson et al. 1999, Siebert et al. 1999). The harbour porpoise is the only cetacean species found on a regular basis in both the German North Sea and Baltic Sea (Reijnders 1992, Benke and Siebert 1994, Schulze 1996, Benke et al. 1998, Hammond et al. 2002). In EU waters, this species is listed in Appendix II of the Bern Convention (implemented in 1982), in Appendix II of the Convention on the Conservation of Migratory Species (CMS; implemented in 1983), in Annex II and IV of the EU Habitats and Species Directive (implemented in 1992), in Annex V of the Convention for the Protection of the Marine Environment of the Northeast Atlantic (Oslo and Paris Convention OSPAR, implemented in 1998), as well as in the German red list of Endangered Species (Boye et al. 1998). The Agreement on the Conservation of Small Cetaceans of the Baltic and

North Seas (ASCOBANS) was concluded in 1991 under the auspices of the CMS (or Bonn Convention) and entered into force in 1994.

Until recently very little data existed on the distribution of harbour porpoises in German waters. Most information on the distribution and population numbers were based on results of the SCANS survey (**S**mall **C**etacean **A**bundance in the **N**orth **S**ea and Adjacent Waters) conducted in July 1994 (Hammond et al. 1995, Hammond et al. 2002). Unfortunately, the coverage during SCANS did not include some areas of the German Exclusive Economic Zone (EEZ), such as the region east of the island of Rügen close to the Polish border in the Baltic Sea, and some parts off the Eastern Friesian Islands, between the estuary of the river Elbe and the Dutch border in the North Sea. In July 2005 SCANS II will take place (http://biology.st-andrews.ac.uk/scans2).

In the process of designating Marine Protected Areas (MPAs) in the framework of the European Habitats Directive (NATURA 2000), the German Federal Agency of Nature Conservation (BfN) selected study areas of particular ecological importance in 2002. This is explained in the introduction and chapter 4 of this book. To evaluate the importance of these sites for harbour porpoises, their distribution and density were studied by conducting aerial surveys in the sites from May 2002 to September 2003. Densities in the study areas were compared between study years and selected areas. The importance of the sites was discussed by taking into account the overall distribution of porpoises in German waters. Answers to the following questions were sought:

- Are harbour porpoises evenly distributed within the study area or is it possible to identify areas of lower or higher density?
- How are the study areas used by porpoises?

2 Methods

2.1 Study Area

Two approaches in survey design were used to answer the above-mentioned questions:

(1) The study area included the EEZ of Germany as well as the 12-nautical mile zone along the coastline of the German North Sea and Baltic Sea (figure 1a). It was divided into 7 substrata (A to G): four located in the North Sea (A–D) and three in the Baltic Sea (E–G). According to their size (A = 3,903 km^2, B = 11,650 km^2, C = 13,668 km^2, D = 11,834 km^2,

E = 4,696 km², F = 7,248 km² and G = 10,990 km²), each region could be surveyed within one day (4–8 hours). In the Baltic Sea, the substrata E and F were extended into Danish waters for logistical reasons so that the northern boundary of the area was determined by the inner Danish islands. These surveys were conducted in the framework of the project MINOS (Marine warm-blooded animals in the North and Baltic Seas: Foundations for assessment of offshore wind farms), funded within the German government's research focus on renewable energies (Investment-in-future program, ZIP, see chapter 14).

(2) The six study areas were surveyed in more detail, that is, more frequently. In the German part of the North Sea these areas were: area 1 = 1,527 km², area 2 = 1,336 km² and area 3 = 5,085 km². In the German Baltic Sea area 4 = 435 km², area 5 = 1,001 km² and area 6 = 3,137 km² were surveyed (figure 1b). These surveys were funded by the BfN.

2.2 Survey Design and Data Acquisition

The surveys followed standard line-transect methodology for aerial surveys (Hiby and Hammond 1989, Buckland et al. 2001). Flights were conducted along a predetermined parallel track design, randomly superimposed on the study area. The direction of transect lines was either north-south or east-west to follow depth gradients (figures 1a and 1b). The aircraft used was a high-winged twin engine Partenavia 68, equipped with bubble windows, flying at an altitude of 183 metres (600 feet) and with a speed of 167 to 186 km/hr (90 to 100 knots). Bubble windows allowed for an unobstructed view on the track. Every four seconds, the aircraft's position was recorded automatically onto a laptop computer connected to a GPS. All sighting positions were stored as well. Sea state (according to the Beaufort scale), glare, cloud cover (parts of eight), turbidity (judged visually: from 0 – clear water with several metres of visibility – to 2 – very turbid water with no visibility under the surface) and sighting probability (judged subjectively as 'good', 'moderate' or 'poor' by observers as the probability of sighting a porpoise given all environmental conditions) were entered at the beginning of each transect and whenever any environmental condition changed. The sighting data was acquired by two observers at the same time, each positioned by a bubble window on both sides of the aircraft. Sighting data included species, declination angle (measured with an inclinometer when the porpoise group was abeam the aircraft), group size, presence of calves, behaviour, swimming direction, cue, reaction to the survey plane, location of porpoise (at surface or under water). A third person,

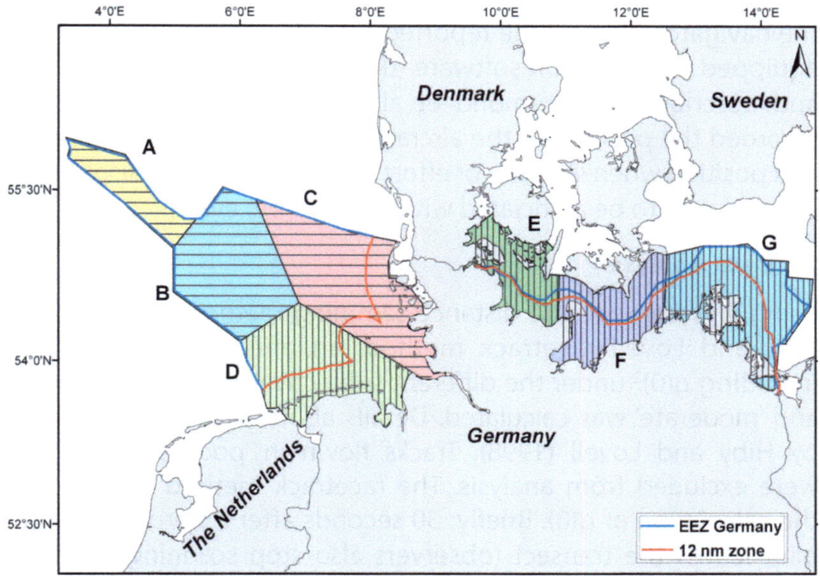

Figure 1a. Study areas of the project MINOS. Transect lines for aerial surveys are indicated by the solid lines. Transect lines are equispaced: 10 km in the North Sea (except area D with 6 km space) and 6 km in all Baltic Sea study areas

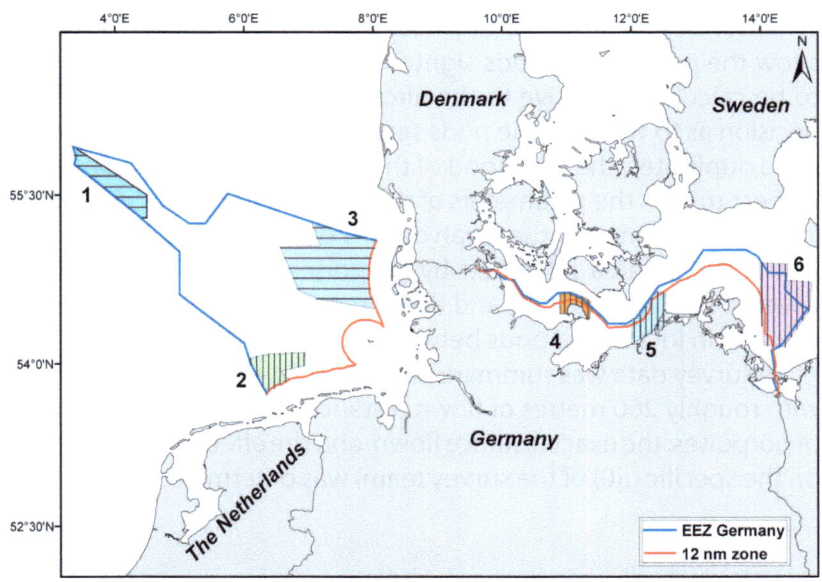

Figure 1b. Intensified study areas in the German EEZ (potential MPAs). Transect lines for aerial surveys are indicated by the solid lines. Transect lines are equispaced: 10 km in the North Sea (except area 2 with 6 km space) and 6 km in all Baltic Sea study areas. 1–*Doggerbank*, 2–*Borkum Reef Ground*, 3–*Sylt Outer Reef*, 4–*Fehmarn Belt*, 5–*Kadet Trench* and 6–*Pommeranian Bay*

the navigator, entered the reported data simultaneously into the laptop equipped with the VOR software (designed by Lex Hiby and Phil Lovell, and described in Hammond et al. 1995). The program continuously recorded the position of the aircraft. The VOR software records the time and position when sighting or effort events occur and it also allows direct entry of data to be associated with the sighting events.

2.3 Data Analysis

Using line-transect and distance sampling methodology as well as the Hiby and Lovell racetrack method, an effective strip width (esw)[1] including g(0)[2] under the different subjective sighting conditions 'good' and 'moderate' was calculated. Details about the method are provided by Hiby and Lovell (1998). Tracks flown in 'poor' sighting condition were excluded from analysis. The racetrack method provides data for the calculation of g(0). Briefly: 30 seconds after a porpoise sighting, the pilot leaves the transect (observers also stop scanning for porpoises), conducts a circle for about 180 seconds and returns to a point in the transect about 30 seconds before the original sighting was made. After being rejoined again with the transect, observers continue searching. Thus, this part of the transect line is surveyed twice (see diagram). The synchronous recording of GPS data, abeam times and declination angles allow the positions of pods sighted on the first and second over-flights to be calculated relative to the aircraft locations at those times. Given a decision as to which of the pods seen on the first and second over-flights were duplicates, the likelihood of those positions can be maximised with respect to g(0), the parameters of the g(y) function and a number of other 'nuisance' parameters: the mean density of porpoise pods in those regions of the survey area inhabited by porpoises, the proportion of the area covered by those regions, and the parameters of the function describing the shift in location of pods between the first and second over-flights.

All survey data was summarised for every 4-second interval, coinciding with roughly 200 metres of flown distance. For each interval the number of porpoises, the exact distance flown, and the effective strip-width (based on the specific g(0) of the survey team) was determined. The distance was

---

[1] esw = the half-strip width of the area searched effectively on each side of the line transect (Buckland et al. 2001).

[2] g(0) = probability of detection on the transect line, usually assumed to be 1. In the case of marine mammals that spend substantial periods underwater and thus avoid detection, this parameter must be estimated from other type of information (Buckland et al. 2001).

multiplied with the total strip-width to obtain the area (in km²) surveyed effectively (i.e., km² on effort). Finally the absolute densities (animals/km²) of harbour porpoises were calculated.

For the purpose of this paper, only summer flights in the period May to August were used for further analysis since the coverage in autumn and winter was very low due to unfavourable weather conditions. Thus, the presented results show the mean summer density of porpoises rather than a snapshot of abundance. Both data sets obtained from May to August 2002 and from May to August 2003 were pooled. For further analysis, the survey area was divided into geographic grid cells of 10 x 10 km (5.4 x 5.4 nautical miles). For each cell the total number of porpoises was divided by the total sum of km² covered on effort. This resulted in density of porpoises per grid cell. Data were analysed and visualised using a Geographical Information System (GIS) software (ArcGIS 8.2).

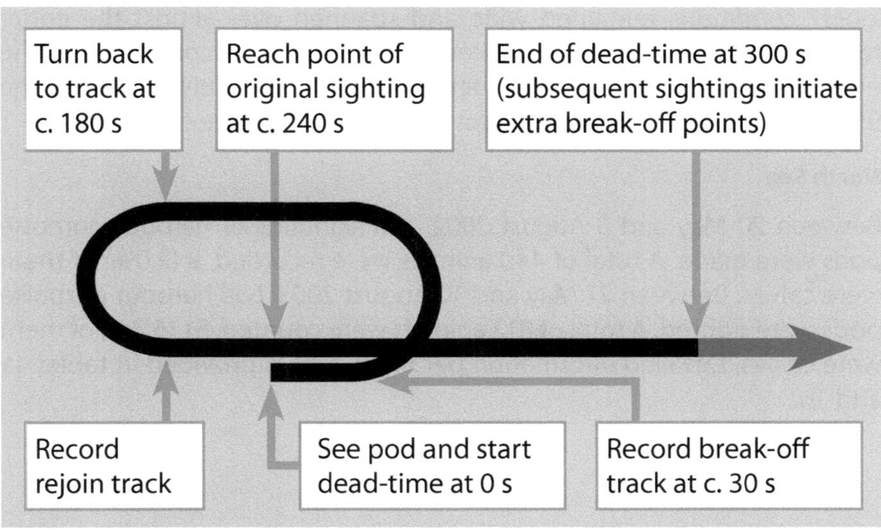

In order to estimate whether densities of harbour porpoises in the same respective areas differed between years and also whether densities differed between areas (in the same respective year), we first determined 95% confidence limits of densities separately for each area and year, based on the track lines covered. These confidence limits estimate an interval that is likely to include the 'true' density. We then checked whether a density of a given area and year fell within the confidence limits of the area and year to which it should be compared. If the density fell outside this range, we concluded that the densities determined for the two areas differed significantly ($p \leq 0.05$) from one another. In order to determine confidence limits, we used a bootstrapping method[3] and determined accelerated bias-corrected 95% confidence limits according to the method described in Manly (1997).

3 Results

3.1 General results

The effective half-strip width calculated, based on the distance of the sightings to the tracklines, was 0.128 km with a g(0) of 0.568 in good conditions. The effective half-strip width was reduced to 0.036 km with a g(0) of 0.164 in moderate conditions. Tracks flown in 'poor' sighting condition were excluded from analysis. Due to the still comparatively low number of racetracks, the 95% confidence limits on g(0) estimated under 'good' conditions remained wide and spanned over almost the entire range from 0 to 1. Additional racetrack flights will be conducted in the near future. The increased number of racetracks is likely to reduce the 95% confidence limit and thus provide a better estimate of g(0).

North Sea

Between 20 May and 3 August 2002, 338 sightings of harbour porpoise pods were made. A total of 440 animals were recorded, 9 (2.0%) of them were calves. Between 27 May and 10 August 2003, 656 harbour porpoise pods were sighted. A total of 812 animals were counted, 51 (6.3%) of them were calves. Detailed information per flight date is provided in tables 1a and 1b.

[3] In principle, this technique is based on repeated random selection of density values determined for single transect lines. The sampling is done with replacement. Once the number of transects sampled equals the number of transects in the study area, the density is determined for the study area as a whole. Confidence limits are then determined by repeating the sampling procedure many times and cutting off the most extreme 5% of the derived distribution of densities. The particular method applied (accelerated bias-corrected confidence limits) corrects for potentially asymmetric distributions. We used 1,000 bootstraps to derive the confidence limit.

Table 1a. Aerial surveys conducted in the German North Sea in 2002

2002	Area	Number of sightings	Number of porpoises	Number of calves	km	km²
20 May	2	6	7	0	54.2	1.9
27, 28 May	3	200	261	3	544.7	83.1
28 May	1	2	4	0	154.4	4.1
4 June	C	13	13	0	420.9	45.6
10 June	C	6	9	0	156.4	1.0
17, 18 June	D	25	42	3	1,203.4	173.3
15 July	C	4	5	0	396.1	21.3
20 July	B	26	31	0	331.6	49.8
20 July	C	12	14	0	117.7	23.9
29 July	C	43	53	3	493.7	53.9
3 August	B	1	1	0	273.2	5.8
Sum		338	440	9	4,146.3	463.7

Table 1b. Aerial surveys conducted in the German North Sea in 2003

2003	Area	Number of sightings	Number of porpoises	Number of calves	km	km²
27 May	3	221	238	4	493.3	103.6
28 May	2	14	14	0	249.7	53.6
30 May	1	38	49	1	167.6	43.0
27 June	C	291	376	29	696.4	141.5
13 July	B	14	18	1	242.1	39.2
31 July	D	2	4	2	249.8	25.5
4 August	A	29	36	2	326.8	62.1
4 August	B	16	25	7	71.8	18.4
7 August	3	31	52	5	430.2	53.0
10 August	2	0	0	0	110.3	7.9
Sum		656	812	51	303.8	547.8

Baltic Sea

Between 18 May and 15 August 2002, 50 sightings of harbour porpoise pods were obtained. A total of 110 animals were counted, one of them (0.9%) was a calf. Between 10 May and 1 August 2003, 34 harbour porpoise pods were sighted. A total of 43 animals were counted, two of them (4.7%) were calves. Detailed information per flight date is provided in tables 2a and 2b.

Table 2a. Aerial surveys conducted in the German Baltic Sea in 2002

2002	Area	Number of sightings	Number of porpoises	Number of calves	km	km^2
18 May	5	0	0	0	165.6	9.1
18 May	6	5	8	0	263.8	47.0
19 May	4	6	9	0	63.7	7.9
12 July	G	32	84	1	834.3	124.8
15 August	F	7	9	0	732.3	105.5
Sum		50	110	1	2,059.7	294.3

Table 2b. Aerial surveys conducted in the German Baltic Sea in 2003

2003	Area	Number of sightings	Number of porpoises	Number of calves	km	km^2
10 May	4	0	0	0	75.2	13.5
14 May	6	0	0	0	158.2	8.1
7 June	G	0	0	0	654.5	133.1
17 June	F	0	0	0	363.9	49.2
17 June	G	0	0	0	107.8	3.6
18 June	F	1	1	0	394.7	46.3
28 June	E	25	30	2	534.7	110.9
1 August	F	8	12	0	438.9	83.2
Sum		34	43	2	2,727.9	447.9

3.2 Distribution of harbour porpoises in German waters

Due to the fact that sighting conditions varied between survey days and areas and sometimes even changed within one day making it often impossible to cover an area in a day, the conducted effort differed between areas (see tables 1 and 2; figures 2a and 3a).

Figure 2b shows harbour porpoise distribution in the **North Sea** study area for the pooled summer flights in 2002 and 2003, respectively. Porpoise density varied over the study area. The north of the survey area showed the highest densities of porpoises. During the flights in May, aggregations of porpoises were seen, indicated by locally high sighting rates of about 40 sightings per 10 km flown distance. Especially the areas off the North Friesian islands of Sylt and Amrum revealed a great abundance of harbour porpoises in the summer months. There seems to be a sharp gradient of density from the northern part to the southern part along the coast. But sighting conditions in the southern

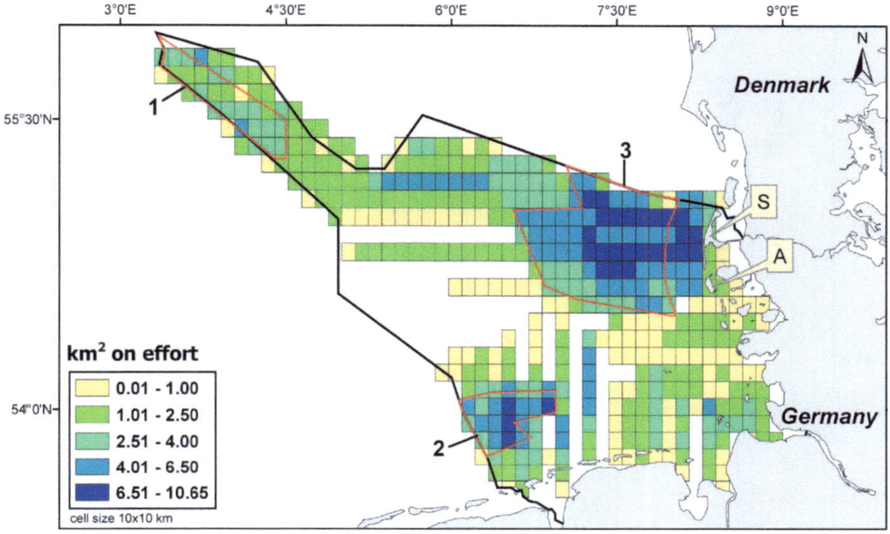

Figure 2a. Km² on effort (i.e., km² surveyed effectively) during the aerial surveys from May to August 2002 and May to August 2003. S=island of Sylt, A=island of Amrum. Map projection: Mercator

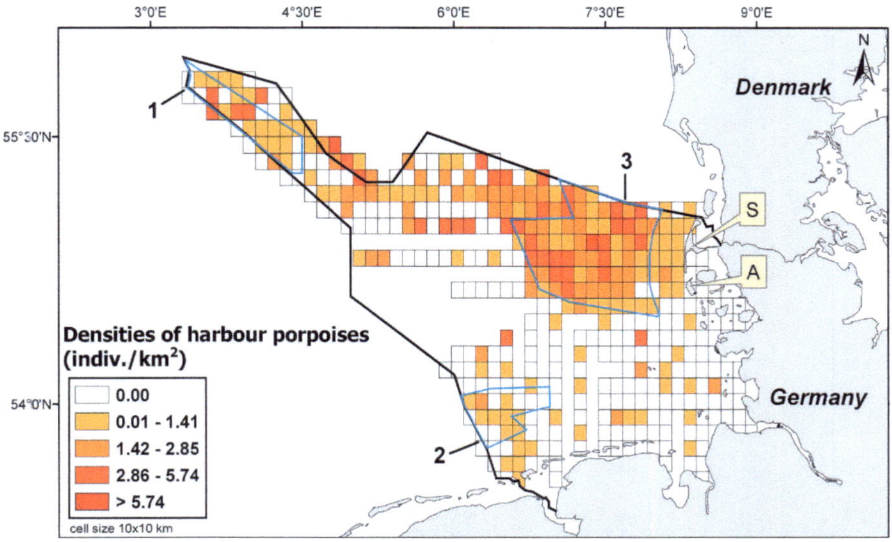

Figure 2b. Summer distribution of harbour porpoises in the German EEZ (black solid line) of the North Sea. All flights conducted in good or moderate conditions between May to August 2002 and May to August 2003 were pooled. S=island of Sylt, A=island of Amrum. Map projection: Mercator

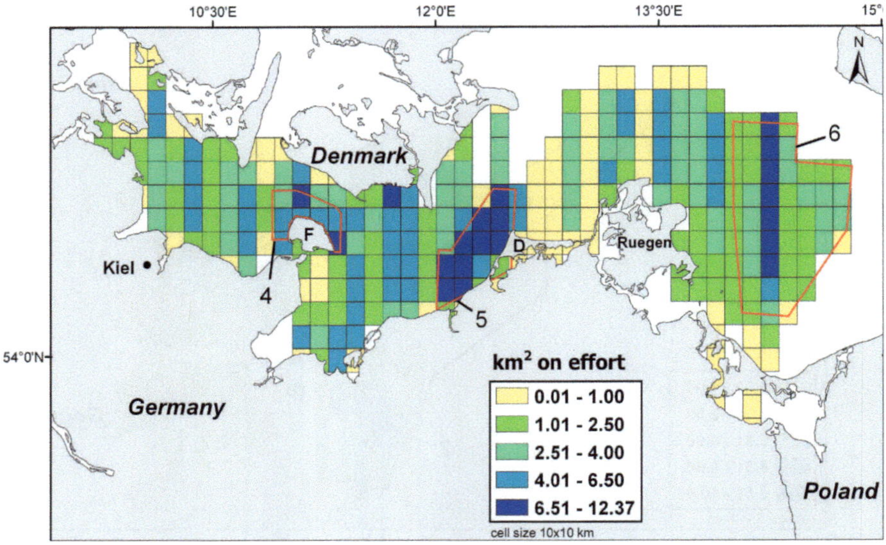

Figure 3a. Km^2 on effort (i.e., km^2 surveyed effectively) during the aerial surveys from May to August 2002 and May to August 2003. F=island of Fehmarn; D=Darss. Map projection: Mercator

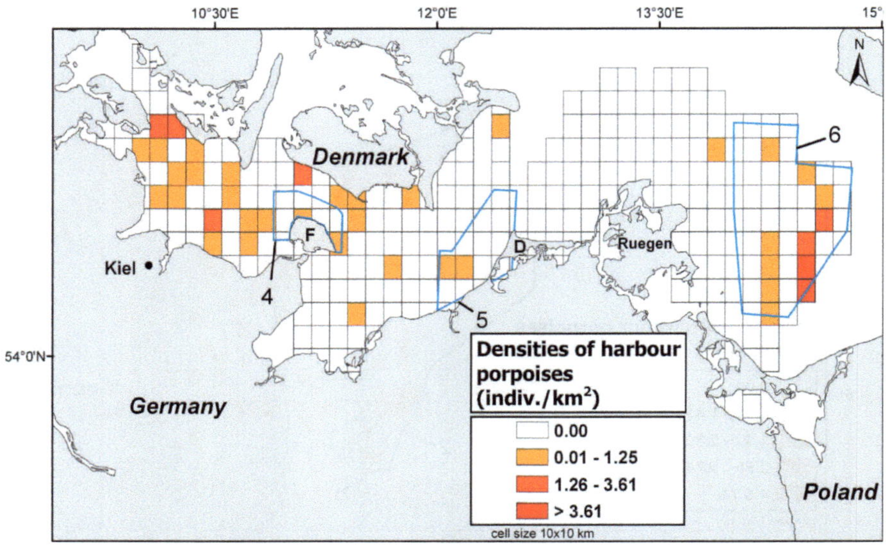

Figure 3b. Summer distribution of harbour porpoises in the Baltic Sea study area. All flights conducted in good or moderate conditions from May to August 2002 and May to August 2003 were pooled. F=island of Fehmarn; D=Darss. Map projection: Mercator

part of substrata B were unfavourable both during 2002 and 2003. Thus, no sightings were obtained in 'good' or 'moderate' conditions.

Harbour porpoise distribution in the **Baltic Sea** is shown in figure 3b. The density of porpoises showed higher values in the western part, namely in the *Kiel Bight* and *Flensburg Fjord*, and in the eastern part close to the border of Poland. But all sightings east of the island of Rügen were only made in 2002. Thus, there is an enormous change in the use of this area between the years. Limited coverage in the western region in 2002 (namely area E) prohibited a direct comparison. Sighting rates were lowest in survey area F. Mean summer density in area E was 0.26 porpoises per km² in 2003. This is higher than the density obtained during SCANS in July 1994 in area X (an area very similar in size and location to area E) with a density of 0.10 porpoises per km² (Hammond et al. 2002). Further information is provided by Scheidat et al. (2004).

3.3 Density estimates for the study areas

In figure 4a the mean summer (May to August) density of harbour porpoises in the three study areas in the **North Sea** is plotted. The highest density in both years was found in area 3 (*Sylt Outer Reef*) with 2.27 animals per km² in 2002, and 2.36 animals per km² in 2003. Lowest densities (0.27 in 2002, and 0.41 in 2003) were calculated for area 2 (*Borkum Reef Ground*). Area 1 (*Doggerbank*) showed a summer density of 0.73 in 2002 and of 0.97 in 2003. 95% confidence limits on these estimates are indicated in figure 4a. They show that the mean density did not differ significantly between years in the same respective area. However, density differed significantly between area 3 and areas 1 and 2, respectively; whereas density difference between areas 1 and 2 was not significant.

The results for the three study areas in the German **Baltic Sea** (figure 4b) are more difficult to interpret. Especially the mean summer density for area 6 (*Pommeranian Bay*) differed strongly between 2002 and 2003. In summer 2002 the density was very high, precisely 0.81 (CL: 0.06–2.04) animals per km², whereas in 2003 no single porpoise was sighted despite high effort. Mean density in area 4 (*Fehmarn Belt*) turned out to be significantly higher in 2002 (0.43) than in 2003 (0.10). In 2002 no harbour porpoise was sighted in area 5 (*Kadet Trench*), whereas in 2003 a mean density of 0.05 was achieved. As the corresponding 95% confidence level span from 0.0 to 0.14 no significant difference was detected between 2002 and 2003 in area 5. The analysis of inter-area specific variation resulted in a significant difference between areas 4 and 5 as well as between areas 5 and 6. No difference could be statistically detected between areas 4 and 6 as confidence levels are very large.

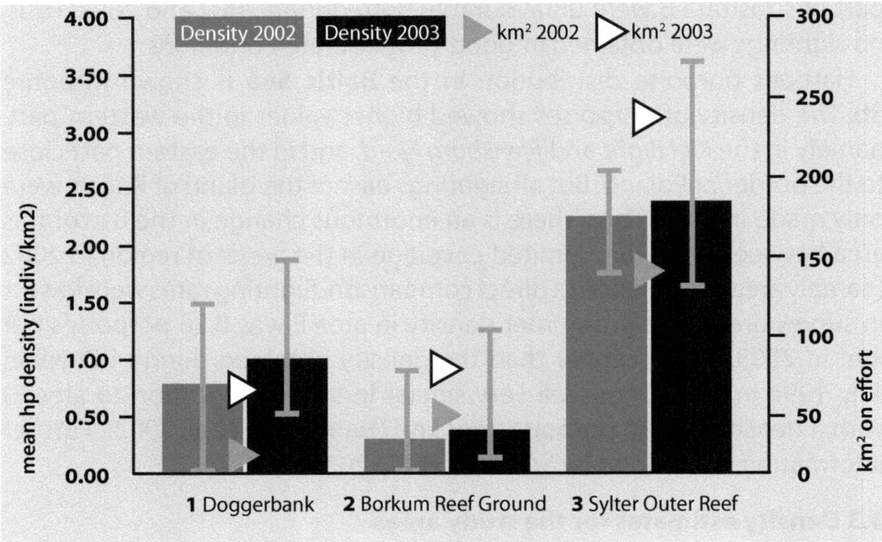

Figure 4a. Mean summer density (left scale) of harbour porpoises in the pSCI study areas of the North Sea (see figure 1b for abbreviations)

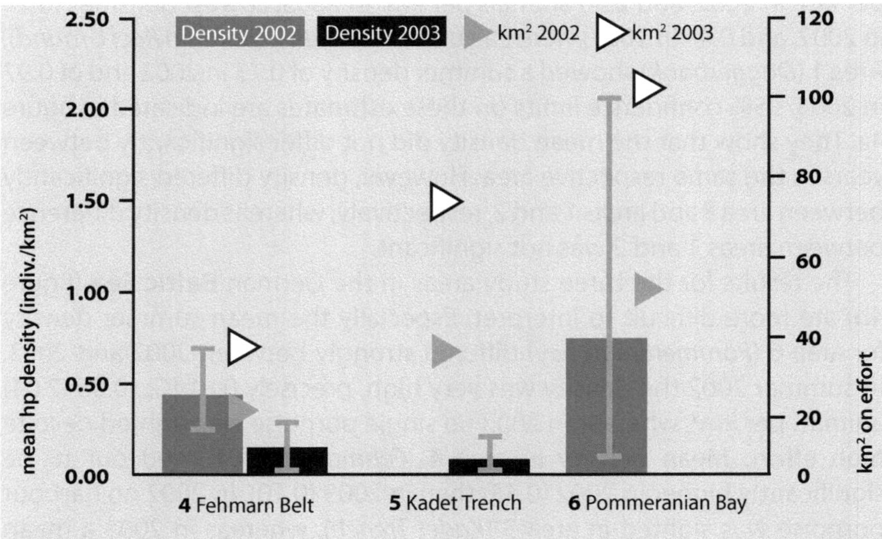

Figure 4b. Mean summer density (left scale) of harbour porpoises in the pSCI study areas of the Baltic Sea (see figure 1b for abbreviations).
Flights conducted in the period May–August 2002 and May–August 2003 were pooled. The upper and lower confidence levels are indicated by the grey line. The tinted and outlined arrowheads show the corresponding effort (right scale, in km²) in 2002 and 2003

4 Discussion

4.1 Distribution patterns and comparison of selected areas

North Sea

The highest number of harbour porpoises was observed in the northern part of the German EEZ. In the remainder of the study area, harbour porpoises were more evenly distributed and no cells with particularly high densities were found. However, coverage under good or moderate conditions was low in the southwestern offshore area, which stresses the importance of conducting further surveys in this area. The high density in the area *Sylt Outer Reef* during the summer survey might be especially related to observed aggregations of animals (i.e., high local sighting rates) in May and June. This seasonal pattern has been observed in both study years around the same time (precisely in 27 and 28 May 2002 and 27 May 2003). The breeding and mating season starts in May (Read 1990, Kinze 1994, Benke et al. 1998). Harbour porpoises might be more gregarious at this time than in other times of the year. The reproductive period is also a life-cycle stage where energy demand is highest (Read 2001). This is especially important for female porpoises as many are simultaneously pregnant and lactating (Read and Hohn 1995, Lockyer et al. 2001). It would therefore be advantageous if lactation occurred when food is abundant and/or of high quality (Börjesson and Read 2003). Pelagic fish, like herring or sprat, have a very high energy content in the summer (Hislop et al. 1991). Swarms of these species might have occurred in the area. Other potential prey species of harbour porpoises are sandeels (*Ammodytes marinus*) which often burrow in the seafloor from October to early April (Wright and Begg 1997, Wright et al. 2000). During late spring and summer they emerge from the seafloor and form dense swarms to feed in the water column (Wright et al. 2000). Distributed in the water column, they are more easily available for predators. Thus, they might be an important food source for porpoises in the North Atlantic (Evans 1990). Analyses of stomach content of porpoises from German waters (1992/1993) showed that 37% of the fish found in the stomachs (by weight) were sandeel. Dab (*Limanda limanda*) and common sole (*Solea vulgaris*) made up 38% and whiting (*Merlangius merlangus*) and cod (*Gadus morhua*) 15% of prey (Benke et al. 1998). If aggregations of harbour porpoise occur due to prey concentrations in certain areas they would most likely occur in late spring and early summer. Similarly, if aggregations occur due to reproductive behaviour these would be expected to be observed in the same time period. A combination of both scenarios is very likely.

Baltic Sea

The population east of the underwater *Darss Sill* (*Darsser Schwelle*) is considered to belong to a different subpopulation than the rest of the Baltic/Belt Sea (e.g., Tiedemann et al. 1996, Börjesson and Berggren 1997, Huggenberger et al. 2002). Joint activities of ASCOBANS and the IWC (International Whaling Commission) have underlined the precarious situation in which this stock seems to be. It seems unlikely that the stock is much larger than 599 animals (CV = 0.57) estimated from a survey in 1995 (Berggren 1995). Therefore, the high densities of porpoises observed in the area *Pommeranian Bay* during the flights in May and June 2002 were quite unexpected. During all other flights sighting rates were extremely low (check areas 6 and G in table 2). Two cruises of the IFAW sailing boat *Song of the Whale*, conducted between Darss ridge and the Bay of Gdansk in Poland in July/August 2001 and 2002 (2,946 km surveyed), have revealed no visual sighting and only three acoustic detections in the area (Gillespie et al. 2002). The most likely explanation for the observed 'hot spots' of porpoises in areas 6 and G in May and July 2002 might be an unusual availability of food. A possible scenario is that porpoises from the Belt Sea, which are part of the subpopulation 'western Baltic' (including Kattegat, Belt Seas, Øresund, *Kiel Bight* and *Fehmarn Belt*), followed their prey into the area of the *Pommeranian Bay*. Again the presence of swarm fish such as herring could also explain the relatively large group sizes. Stomach analyses of stranded harbour porpoises along the German coast of the Baltic showed that 22.8% of the fish found (by weight) was herring, 52.7% goby and 14.8% cod (Benke et al. 1998). Large aggregations of up to several hundred harbour porpoises have been observed in other areas of the world, probably related to good feeding grounds (Rae 1965, Evans 1990). If valuable prey is only available for a short period of time, such as spawning shoals of herring or sprat, these aggregations of harbour porpoises might be difficult to encounter using standard line-transect methodology in such a low density area like the eastern Baltic Sea.

A general west–east gradient in harbour porpoise densities is very likely. The high density in area E and the gradual decline in density while moving to the east (e.g., areas F: 0.06 in 2003 and G: 0 porpoises per km^2 in 2003) underline that theory. A robust analysis of the results for the study areas *Fehmarn Belt* and *Kadet Trench* is difficult as these survey areas were very small in size. Thus, detecting a sufficient number of sightings for robust statistics was impossible. The mean length of the six transects in area 4, for example, was only 14 km. An aircraft flying at 100 knots covers this distance in 4 to 5 minutes.

5 Conclusion

The 2002 and 2003 aerial surveys in German waters revealed new and, in some respects, unexpected information on distribution of porpoises. The results allowed us to answer the main question of this study.

- **Are porpoises evenly distributed within the study area or can we identify areas of lower or higher density?**

The main results showed large aggregations and high densities of porpoises in the north-eastern part of the German EEZ in the North Sea. Especially the areas off the North Friesian islands of Sylt and Amrum revealed a great abundance of harbour porpoises in the summer months. There seems to be a sharp gradient of density running from north to south. In the eastern part of the Baltic Sea high densities were observed in summer 2002. As subsequent surveys did not yield a single sighting, an explanation for this phenomenon remains speculative. Besides that, a clear west–east gradient in harbour porpoise density was ascertained. Further large scale information on abundance, distribution and stock identities are necessary to put into a broader context the observations from this study.

- **How are the pSCI (proposed Sites of Community Importance) used by porpoises?**

The sites were used differently by harbour porpoises. Within the North Sea the highest density was found in the area *Sylt Outer Reef*. This was consistent during the two survey years. Our results clearly indicate that this site is very important during the sensitive reproductive period. The offshore area *Doggerbank* was only covered twice by flights due to logistical difficulties associated with flying in offshore areas. The densities estimated for this site were fairly high indicating an important area for porpoises. The lowest densities in the study areas of the North Sea were found in the area *Borkum Reef Ground*. Generally the high confidence limits of the density estimates of both *Doggerbank* and *Borkum* are related to the smaller size of the areas and the lower sighting rate, thus making it difficult to evaluate their importance.

In the Baltic Sea highest densities were found in the area *Pommeranian Bay* in 2002. However, no sightings were made in this area in 2003. Continuing surveys are carried out to determine how regularly this area is used by harbour porpoises and again in early summer 2005 some individuals were sighted. The aerial surveys show that the remaining two study areas *Fehmarn Belt* and *Kadet Trench* are used by porpoises, especially the area around the island of Fehmarn. For a detailed monitoring of how porpoises use those fairly small areas the use of stationary acoustics (see chapter 12) are applied as well.

Acknowledgements

This work was made possible due to two projects which investigate the distribution and abundance of marine mammals in German waters. The German Federal Ministry for the Environment, Nature Conservation and Nuclear Safety (BMU) is funding within the Investment-in-future program (ZIP), the project MINOS (Marine Warmblüter in Nord- und Ostsee), and the project EMSON (Erfassung von Meeressäugetieren und Seevögeln in der deutschen AWZ der Nord- und Ostsee), the latter being under the management of the Federal Agency for Nature Conservation (BfN). We would like to thank Lex Hiby and Phil Lovell for the calculation of strip-width and g(0). Our special thanks go to the pilots of the survey planes, especially Peter Siemiatkowski from Sylt Air and Leif Petersen from the Danish Air Survey. The completion of the surveys was only possible through the dedication of the observers Jörg Adams, Patrik Börjesson, Helena Herr, Iwona Kuklik, Kristina Lehnert, Maik Marahrens and Carsten Rocholl. Data analyses were only possible due to the help of Hauke Giewat, Ulrike Kleeberg and Roger Mundry.

References

Ascobans (2000) Agreement on the Conservation of Small Cetaceans of the Baltic and North Seas. Proceedings of the third meeting of parties to ASCOBANS. Bristol, UK, 26–28 July 2000

Benke H, Siebert U (1994) Zur Situation der Kleinwale im Wattenmeer und in der südlichen Nordsee. In: Lozán JL, Rachor E, Reise K, v. Westernhagen H, Lenz W (eds) Warnsignale aus dem Wattenmeer. Blackwell Wissenschaftsverlag, Berlin, pp 309–316

Benke H, Siebert U, Lick R, Bandomir B, Weiss R (1998) The current status of harbour porpoises (*Phocoena phocoena*) in German waters. Archive for Fishery and Marine Research 46 (2):97–123

Berggren P (1995) A preliminary assessment of the status of harbour porpoises (*Phocoena phocoena*) in the Swedish Skagerrak, Kattegat and Baltic seas. Paper SC/47/SM50 presented to the IWC Scientific Committee, May 1995 (unpublished)

Börjesson P, Berggren P (1997) Morphometric comparisons of skulls of harbour porpoises (*Phocoena phocoena*) from the Baltic, Kattegat and Skagerrak seas. Canadian Journal of Zoology 75:280–287

Börjesson P, Read AJ (2003) Variation in timing of conception between populations of the harbour porpoise. Journal of Mammalogy 84:948–955

Boye P, Hutterer R, Benke H (1998) Rote Liste der Säugetiere (Mammalia). In: Bundesamt für Naturschutz (eds) Rote Liste gefährdeter Tiere Deutschlands. Münster (Landwirtschaftsverlag). Schriftenreihe für Landschaftspflege und Naturschutz 55:33–39

Buckland ST, Anderson DR, Burnham KP, Laake J, Thomas L (2001) Introduction to Distance Sampling: Estimating abundance of biological populations. Oxford University Press, New York

Evans PGH (1990) European cetaceans and seabirds in an oceanographic context. Lutra 33 (2):95–125

Gillespie D, Brown S, Lewis T, Matthews J, McLanaghan R, Moscrop A (2002) Preliminary results from acoustic and visual surveys for harbour porpoises (*Phocoena phocoena*) in German, Danish, Swedish and Polish waters during summer 2002. IFAW 'Quick Look' Report. Song of the Whale Research Team, International Fund for Animal Welfare, London, UK (unpublished)

Hammond PS, Benke H, Berggren P, Borchers DL, Buckland ST, Collet A, Heide-Jørgensen MP, Heimlich Boran S, Hiby AR, Leopold MF, Øien N (1995) Distribution and abundance of harbour porpoises and other small cetaceans in the North Sea and adjacent waters. European Community LIFE Programme, Life, LIFE 92-2/IL/027

Hammond PS, Berggren P, Benke H, Borchers DL, Buckland ST, Collet A, Heide-Jørgensen MP, Heimlich S, Hiby AR, Leopold MF, Øien N (2002) Abundance of harbour porpoise and other cetaceans in the North Sea and adjacent waters. Journal of Applied Ecology 39:361–376

Hiby AR, Hammond PS (1989) Survey techniques for estimating the abundance of cetaceans. Reports of the International Whaling Commission, Special Issue 11:47–80

Hiby AR, Lovell P (1998) Using aircraft in tandem formation to estimate abundance of harbour porpoise. Biometrics 54:1280–1289

Hislop JRG, Harris MP, Smith JGM (1991) Variation in the calorific value and total energy content of the lesser sandeel (*Ammodytes marinus*) and other fish preyed on by seabirds. Journal of Zoology 224:501–517

Huggenberger S, Benke H, Kinze CC (2002) Geographical variation in harbour porpoise (*Phocoena phocoena*) skulls: support for a separate non-migratory population in the Baltic Proper. Ophelia 56 (1):1–12

Jepson PD, Baker JR, Allchin CR, Law RJ, Kuiken T, Baker JR, Rogan E, Kirkwood JK (1999) Investigating potential associations between chronic exposure to polychlorinated biphenyls and infectious disease mortality in harbour porpoises from England and Wales. The Science of the Total Environment 243/244:339–348

Kinze CC (1994) *Phocoena phocoena* (Linnaeus 1758) – Schweinswal oder Kleintümmler (auch Braunfisch). In: Niethammer J, Krapp F (eds) Handbuch der Säugetiere Europas. Band 6: Meeressäuger. Aula Verlag, Wiesbaden, pp 242–264

Lockyer C, Heide-Jørgensen MP, Jensen J, Kinze CC. Sørensen TB (2001) Age, length and reproductive parameters of harbour porpoises *Phocoena phocoena* (L.) from West Greenland. ICES Journal of Marine Science 58:154–162

Manly BFJ (1997) Randomization, Bootstrap and Monte Carlo Methods in Biology. Chapman & Hall, New York

Rae BB (1965) The food of the common porpoise (*Phocoena phocoena*). Journal of Zoology London 146:114–122

Read AJ (1990) Reproductive seasonality of harbour porpoises from the Bay of Fundy. Canadian Journal of Zoology 68:284–288

Read AJ (2001) Trends in the maternal investment of harbour porpoises are uncoupled from the dynamics of their primary prey. Proceedings of the Royal Society of London B 268:573–577

Read AJ, Hohn AA (1995) Life in the fast lane: The life history of harbor porpoises from the Gulf of Maine. Marine Mammal Science 11:423–440

Reijnders PJH (1992) Harbour porpoises *Phocoena phocoena* in the North Sea: Numerical responses to changes in environmental conditions. Netherlands Journal of Aquatic Ecology 26(1):75–85

Scheidat M, Kock K-H, Siebert U (2004) Summer distribution of harbour porpoise (*Phocoena phocoena*) in the German North and Baltic Sea. Journal of Cetacean Research and Management 6 (3):251-257

Schulze G (1996) Die Schweinswale. Neue Brehm Bücherei, Magdeburg

Siebert U, Joiris C, Holsbeek L, Benke H, Failing K, Frese K, Petzinger E (1999) Potential relation between mercury concentrations and necropsy findings in cetaceans from German waters of the North and Baltic Seas. Marine Pollution Bulletin 38:285–295

Tiedemann R, Harder J, Gmeiner C, Haase E (1996). Mitochondrial DNA patterns of harbour porpoises (*Phocoena phocoena*) from the North and the Baltic Sea. Zeitschrift für Säugetierkunde 61:104–111

Vinther M (1999) Bycatches of harbour porpoises (*Phocoena phocoena* L.) in Danish set-net fisheries. Journal of Cetacean Research and Management 1 (2):123–135

Wright PJ, Begg GS (1997) A spatial comparison of common guillemots and sandeels in Scottish waters. ICES Journal of Marine Science 54:578–592

Wright PJ, Jensen H, Tuck I (2000) The influence of sediment type on the distribution of the lesser sandeel, *Ammodytes marinus*. Journal of Sea Research 44:243–256

Chapter 12

Seasonal and geographical variation of harbour porpoise (*Phocoena phocoena*) habitat use in the German Baltic Sea monitored by passive acoustic methods (PODs)

Ursula K. Verfuß, Christopher G. Honnef and Harald Benke

German Oceanographic Museum, Stralsund

Abstract

Harbour porpoises (*Phocoena phocoena*) were known to be common in the Baltic Sea. In the past several decades, the abundance and distribution has decreased, leading to national and international agreements on the protection of this species. Plans for offshore windmill constructions and proposals for Marine Protected Areas (MPAs) to implement NATURA 2000, led to an increased research effort on the harbour porpoise in the German Exclusive Economic Zones (EEZs) of the North and Baltic Sea. Within this scope, the harbour porpoise habitat use of the German Baltic Sea from Fehmarn to the *Pommeranian Bay* (*Pommersche Bucht*) was investigated with the help of self-contained submersible data logger (Porpoise detectors, T-PODs), which register harbour porpoise echolocation click trains.

Comparison of the T-POD data from different measuring stations located throughout the Baltic Sea revealed a decrease of porpoise registrations from the west of the island of Fehmarn to the east of the island of Rügen. Seasonal variation of habitat use, and therefore of relative porpoise density, was seen around the island of Fehmarn and the *Kadet Trench* (*Kadetrinne*), with many days of porpoise registrations in the summer and fewer registration days in the winter months.

The results prove the regular use of the western part of the German EEZ of the Baltic Sea by harbour porpoises from Fehmarn to the *Kadet Trench* including adjacent coastal waters. The low amount of porpoise registrations east of the *Darss Sill* (*Darsser Schwelle*) allows the assumption of a low harbour porpoise density in the eastern part of the German Baltic

Sea. Furthermore, a clear seasonal variation in the amount of porpoise registration proves porpoise migration out of the western part of the German Baltic Sea in wintertime.

1 Introduction

Harbour porpoises (*Phocoena phocoena*) have been very common in the North Sea and Baltic Sea up to the middle of the 20th century (Schulze 1996). In the past several decades, a drastic decrease in their population size – as indicated for some areas (Benke and Siebert 1994, Kinze 1995, Kröger 1986, Reijnders 1992, Siebert et al. 1996) – has led to the endangerment of the porpoise population (e.g., ICES/ACME 1997). Harbour porpoises are now protected by a variety of national and international laws and agreements: ASCOBANS, HELCOM, OSPARCOM, and Red list of mammals, Germany (Boye et al. 1998).

Former research on abundance and distribution of harbour porpoises (Benke et al. 1998, Hammond et al. 2002, Heide Jørgensen et al. 1993, Sonntag et al. 1999) gave neither a complete picture of the distribution pattern nor any information on seasonality in the German Baltic Sea. Therefore, plans for constructing offshore wind farms and proposals for MPAs to implement the European habitat directive NATURA 2000 led to an increased research effort on the harbour porpoise in Germany. Recent aerial surveys investigated the spatial distribution of harbour porpoises in the German part of the North and Baltic Sea (see chapter 11). Parallel to this, passive acoustic monitoring devices T-PODs (Porpoise Detectors) were deployed permanently on measuring positions throughout the German Baltic Sea from Fehmarn to the *Pommeranian Bay*.

The harbour porpoise, like other odontocete species, emit short-pulsed high frequency click sounds for echolocation (Au 1993). As an active sensory system, echolocation in porpoises is used for orientation as well as for foraging (Verfuß and Schnitzler 2002). Harbour porpoise echolocation clicks are very distinct and different from most dolphin echolocation clicks (Au 1993). Their main energy is focused on a small frequency bandwidth around 130 kHz (Goodson et al. 1995, Kamminga et al. 1999). The method of passive acoustic monitoring with T-PODs takes advantage of the highly specialised sonar system of porpoises. The distinct and easily distinguishable click structure provides a good opportunity to set up an automatic system that specifically monitors this species.

The advantage of this kind of acoustic monitoring is that, in contrast to aerial surveys which use snapshots of harbour porpoise sightings to determine distribution and abundance, T-PODs are for long-term deployment. They register the presence of harbour porpoises over months.

This paper presents the results of the first year (August 2002 to August 2003) of continuous monitoring for harbour porpoise echolocation activity in most parts of the German Baltic Sea.

2 Methods

2.1 Methodology

T-PODs are self-contained data loggers for cetacean echolocation clicks (for details, see www.chelonia.demon.co.uk/PODhome.html), consisting of a hydrophone, filter, and memory (figure 1). They register, in a 10-µsec resolution, the presence and length of high frequency click sounds matching specific criteria, logging for 24 hours a day for a period of eight to ten weeks. After this period, the data are downloaded and batteries have to be replaced.

Figure 1. A T-POD moored under water

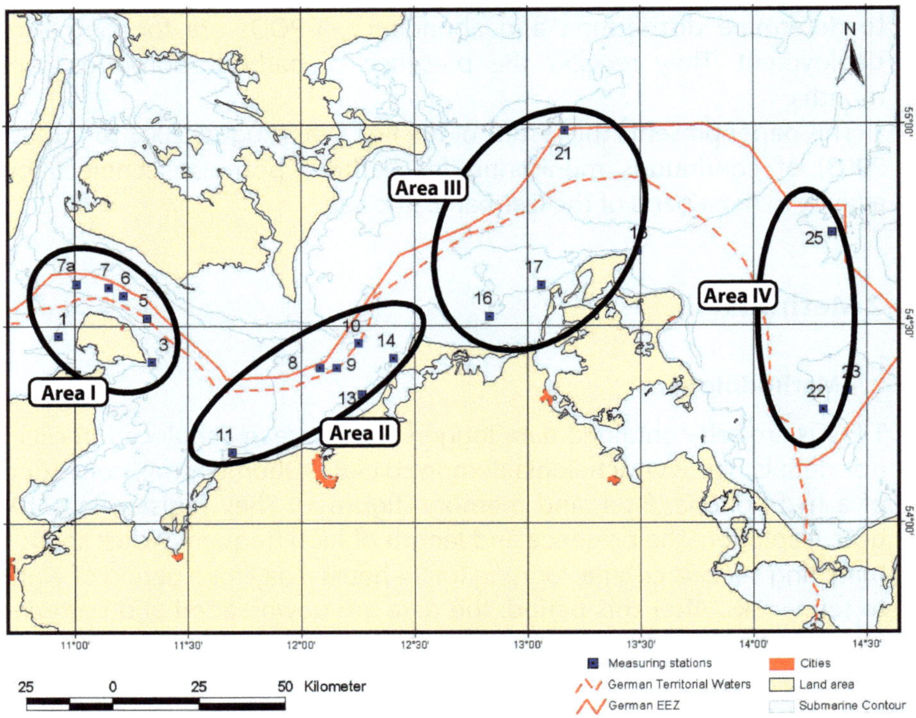

Figure 2. Locations of all utilised T-POD measuring stations in the Baltic Sea. The area of investigation was divided into four sub-areas (black circles): area I: stations 1, 3, 5–7a; area II: stations 8–11, 13, 14; area III: stations 16–18, 21; area IV: stations 22, 23, 25

2.2 T-POD application

Nineteen (19) measuring positions were selected to monitor the German Baltic Sea from Fehmarn to the *Pommeranian Bay* (figure 2). On each measuring position, one T-POD at a time was deployed on a mooring, fixed five to seven metres under the water surface. T-PODs of versions 2 and 3 were used. The mooring consisted of a 30-kg anchor connected to several surface buoys via a rope (figure 3).

The listening criteria of the T-PODs were set to "porpoise-only high sensitivity" as given in the T-POD programme (T-POD version 2: filter A = 130 kHz, filter B = 90 kHz, ratio A/B = 4, 'A' filter sharpness = 10, 'B' filter sharpness = 18, minimum intensity = 6, scan limit on number (N) of clicks logged = 240; T-POD version 3: filter A = 130 kHz, filter B = 90 kHz, ratio A/B = 4, 'A' integration period = short, 'B' integration period = long, minimum

Figure 3. Surface markers of a T-POD mooring in the Baltic Sea

intensity = 6, scan limit on N clicks logged = 240). Where background noise did not allow these settings, the ratio A/B was set to 6, which reduced the registration of high frequency background noise. This change in the settings affected neither the sensitivity nor the comparability of the gathered data (Verfuß et al. 2004a). Data recorded with version 2 T-PODs were comparable with the data of version 3 T-PODs (Verfuß et al. 2004b).

The T-PODs were calibrated before deployment to determine the minimum receiving level of each T-POD. This is the level at which the device will start to register porpoise clicks. The minimum receiving level of the deployed T-PODs was in the range of 117 dB re 1 $V_{(pp)}$/µPa up to 144 dB re 1 $V_{(pp)}$/µPa. Lower receiving level means a more sensitive T-POD and vice versa.

2.3 Data analysis

The click sounds registered from the T-PODs were scanned for trains of clicks with a specific signal pattern by means of a Train Detection algorithm (V2.2), which was included in the T-POD software. Click trains classified by the algorithm as "high probability cetacean click trains" up to "very doubtful trains" originated from harbour porpoises, boat noise (e.g., sonar, propeller noise), or background noise. Those click trains were manually reviewed for harbour porpoise echolocation click trains as described in Verfuß et al. (2004a, 2004b). Click trains classified by the algorithm, and which were then manually attributed to porpoise origin, were included in the data set. Those that were manually attributed to other sources were rejected.

For further analysis, porpoise-positive days, defined as a day with at least one classified porpoise click train, were determined from all data recordings. The percentage of porpoise-positive days in the number of monitored days per month was calculated for each position. Months with less than five monitoring days were ignored.

The monitored area of the German Baltic Sea was divided into four sections each with the following T-POD positions:

- Area I: positions 1 to 7: western part of the German Baltic Sea, area around Fehmarn island
- Area II: positions 8 to 14: western part of the German Baltic Sea, *Kadet Trench* and adjacent coastal area
- Area III: positions 16 to 21: eastern part of the German Baltic Sea, area north of Darss and around Rügen island including EEZ
- Area IV: positions 22 to 25: eastern part of the German Baltic Sea, *Pommeranian Bay*

The mean of the percentages of porpoise-positive days per month from the included positions was calculated for each of the four areas.

2.4 Influence of T-POD sensitivity

An earlier work by the present authors (Verfuß et al. 2004b) showed that the difference in T-POD sensitivity of applied T-PODs could have an influence in the comparability of gathered data. Several T-PODs were simultaneously deployed in an area with high porpoise abundance. Those T-PODs had a range of sensitivities, comparable to the sensitivity range of the T-PODs used in this investigation. Analysis of porpoise-positive hours (i.e., hours with at least one porpoise registration), revealed a linear relationship between this parameter and the T-POD sensitivity. It was concluded that using the parameter of porpoise-positive days will have less influence on the data comparability since it does not depend on the amount of porpoise-positive hours whether a day is porpoise-positive or not.

To test for differences between areas while simultaneously controlling for T-POD sensitivity, an ANCOVA[1] was used. Since the data were not normally distributed, we applied this model to the original data as well as to the ranked data. The interaction between the covariate (T-POD sensitivity) and the factor (area) was initially included into the model. Since in both analyses the interaction term was not significant ($F_{3,107} \leq 1.498$, $P \geq 0.219$), we removed the interaction term from the model in both cases. Here we only report the results with the interaction term removed.

[1] ANCOVA: Statistical analysis of covariance which simultaneously considers the effect of two independent variables – one varying categorical, and the other one varying continuously – on a dependent variable (simplest model).

3 Results

Table 1 (overleaf) shows an overview of the number of monitored days per month and the corresponding percentage of porpoise-positive days per month for each T-POD position, as well as the sensitivity of the applied T-POD during the specific month. None of the positions were monitored for the entire time due to logistical reasons and to loss of moorings, in some cases. The total amount of observed days is indicated.

We found no significant relation between T-POD sensitivity and the percentage of porpoise-positive days per month (unranked data: $F_{1,110} = 0.488$, $P = 0.486$; ranked data: $F_{1,110} = 0.038$, $P = 0.846$). Areas clearly differed in the percentage of porpoise-positive days per month (unranked data: $F_{3,110} = 92.263$, $P<0.001$; ranked data: $F_{3,110} = 69.475$, $P<0.001$).

The results show a geographical as well as a seasonal variation in the percentage of porpoise-positive days from the total number of days on which data were obtained (figure 4):

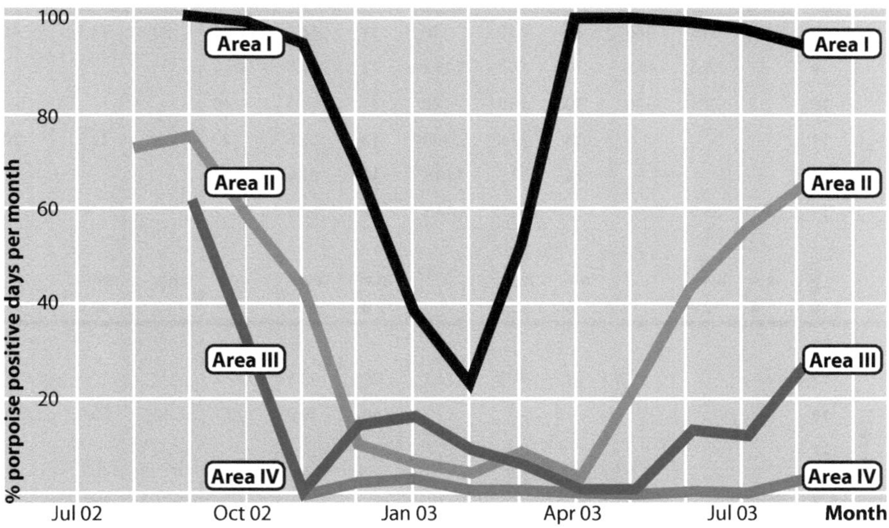

Figure 4. Mean percentage of porpoise-positive days per month for area I to area IV over a one-year period (August 2002 to August 2003). Measuring stations included: area I: stations 1, 3, 5–7a; area II: stations 8–11, 13, 14; area III: stations 16–18, 21; area IV: stations 22, 23, 25.

Table 1. An overview of the number of monitored days per month and the corresponding percentage of porpoise-positive days per month for each T-POD position, as well as the sensitivity of the applied T-POD during the specific month

		2002											
		Aug			Sep			Oct			Nov		
		obs days	pp days (%)	T-POD-sensitivity	obs days	pp days (%)	T-POD-sensitivity	obs days	pp days (%)	T-POD-sensitivity	obs days	pp days (%)	T-POD-sensitivity
Area I	1							9	100.0	119.3	30	96.7	119.3
	3							9	100.0	117	30	90.0	117
	5							23	95.6	121	18	100.0	121
	6										18	94.4	121.7
	7a												
	7				19	100.0	118.1	17	100.0	118.1			
	n/aver.	n = 0			n = 1	aver. 100.0		n = 4	aver. 98.9		n = 4	aver. 95.3	
Area II	8	31	64.5	146	30	80.0	146	31	58.1	146	30	73.3	146
	9	31	64.5	144.2	30	73.3	144.2	22	45.5	144.2			
	10	31	77.4	140	30	66.7	140	31	64.5	140			
	11				18	77.8	120	19	52.6	120	8	12.5	120
	13				18	83.3	119.1	16	62.6	119.1			
	14	31	83.9	127.7	15	72.0	124.95	16	62.5	122.2			
	n/aver.	n = 4	aver. 72.6		n = 6	aver. 75.5		n = 6	aver. 57.6		n = 2	aver. 42.9	
Area III	16												
	17				13	61.5	127.7	30	33.3	127.7			
	18							10	30.0	123.1	30	0.0	123.1
	21												
	n/aver.	n = 0			n = 1	aver. 61.5		n = 2	aver. 31.7		n = 1	aver. 0.0	
Area IV	22												
	23										15	0.0	117.2
	25										16	0.0	127.7
	n/aver.	n = 0			n = 0			n = 0			n = 2	aver. 0.0	

Note
n = total number of T-PODs applied in each area during a specific month
aver. = average percentage of porpoise-positive (pp) days per month in a specific area

		2002 continued					2003					
		Dec			Total	Total	Jan			Feb		
		obs days	pp days (%)	T-POD-sensitivity	2002 obs days total	2002 pp days total (%)	obs days	pp days (%)	T-POD-sensitivity	obs days	pp days (%)	T-POD-sensitivity
Area I	**1**	31	41.9	119.3	**70**	**79.5**	13	38.5	119.3	26	23.1	117
	3	17	52.9	117	**56**	**81.0**						
	5	18	88.9	121	**59**	**94.8**						
	6	18	94.4	121.7	**36**	**94.4**						
	7a											
	7				**36**	**100.0**						
	n/aver.	**n =** **4**	aver. 69.6		**n =** **5**	aver. 90.0	**n =** **1**	aver. 38.5		**n =** **1**	aver. 23.1	
Area II	**8**	31	19.4	146	**153**	**59.1**	31	3.2	146	28	3.6	146
	9				**83**	**61.1**						
	10				**96**	**69.5**						
	11	31	0.0	120	**76**	**35.7**	11	9.1	120			
	13				**34**	**72.9**						
	14				**66**	**72.8**						
	n/aver.	**n =** **2**	aver. 9.7		**n =** **6**	aver. 61.9	**n =** **2**	aver. 6.2		**n =** **1**	aver. 3.6	
Area III	**16**											
	17				**43**	**47.4**						
	18	31	22.6	123.1	**71**	**17.5**	31	9.7	123.1	28	3.6	123.1
	21	19	5.3	128.9	**19**	**5.3**	31	22.6	128.9	28	14.3	128.9
	n/aver.	**n =** **2**	aver. 13.9		**n =** **3**	aver. 23.4	**n =** **2**	aver. 16.1		**n =** **2**	aver. 8.9	
Area IV	**22**											
	23	26	0.0	130.7	**41**	**0.0**	31	0.0	144.2	28	0.0	144.2
	25	31	3.2	127.7	**47**	**1.6**	31	6.5	127.7	28	0.0	127.7
	n/aver.	**n =** **2**	aver. 1.6		**n =** **2**	aver. 0.8	**n =** **2**	aver. 3.2		**n =** **2**	aver. 0.0	

Table 1 continued. An overview of the number of monitored days per month and the corresponding percentage of porpoise-positive days per month for each T-POD position, as well as the sensitivity of the applied T-POD during the specific month

		2003 continued								
		Mar			Apr			May		
		obs days	pp days (%)	T-POD-sensitivity	obs days	pp days (%)	T-POD-sensitivity	obs days	pp days (%)	T-POD-sensitivity
Area I	1	31	51.6	117	30	100.0	117	31	100.0	117
	3									
	5				16	100.0	121	31	100.0	121
	6				16	100.0	118.1	31	100.0	119.9
	7a				16	100.0	125	31	100.0	125
	7									
	n/aver.	n = **1**	aver. 51.6		n = **4**	aver. 100.0		n = **4**	aver. 100.0	
Area II	8	12	16.7	146						
	9									
	10	14	7.1	126.2	30	6.7	126.2	31	12.9	126.2
	11									
	13									
	14	14	0.0	126.9	30	0.0	126.9	16	31.3	126.9
	n/aver.	n = **3**	aver. 7.9		n = **2**	aver. 3.3		n = **2**	aver. 22.1	
Area III	16							6	0.0	118.1
	17				5	0.0	129	31	0.0	129
	18	18	11.1	123.1	30	0.0	123.1	31	3.2	123.1
	21	10	0.0	128.9	30	0.0	128.9	31	0.0	128.9
	n/aver.	n = **2**	aver. 5.5		n = **3**	aver. 0.0		n = **4**	aver. 0.8	
Area IV	22									
	23	31	0.0	144.2	30	0.0	144.2	31	0.0	144.2
	25	31	0.0	127.7	30	0.0%	127.7	31	0.0	127.7
	n/aver.	n = **2**	aver. 0.0		n = **2**	aver. 0.0		n = **2**	aver. 0.0	

Note
n = total number of T-PODs applied in each area during a specific month
aver. = average percentage of porpoise-positive (pp) days per month in a specific area

		2003 continued										
		Jun			**Jul**			**Aug**			**Total**	**Total**
		obs days	**pp days (%)**	**T-POD-sensitivity**	**obs days**	**pp days (%)**	**T-POD-sensitivity**	**obs days**	**pp days (%)**	**T-POD -sensitivity**	**2003 obs days total**	**2003 pp days total (%)**
Area I	**1**	20	100.0	117	14	100.0	125	31	100.0	125	**196**	**76.6**
	3											
	5	30	100.0	121	31	96.8	121	31	90.3	121	**139**	**97.4**
	6	30	100.0	121.7				17	94.1	126	**96**	**98.5**
	7a	30	96.7	125							**79**	**98.9**
	7											
	n/aver.	n = **4**	aver. 99.2		n = **2**	aver. 98.4		n = **3**	aver. 94.8		n = **4**	aver. 75.7
Area II	**8**				11	54.5	126.2	31	48.4	126.2	**113**	**25.3**
	9											
	10	30	43.3	126.2	31	61.3	126.2	31	77.4	126.2	**167**	**34.8**
	11				11	36.4	126.9	15	53.3	126.9	**37**	**32.9**
	13											
	14				17	70.6	121.7	23	82.6	121.7	**100**	**36.9**
	n/aver.	n = **1**	aver. 43.3		n = **4**	aver. 55.7		n = **4**	aver. 65.4		n = **4**	aver. 32.5
Area III	**16**	30	23.3	118.1	31	29.0	118.1	27	40.7	118.1	**94**	**23.3**
	17	30	13.3	129	20	10.0	129	28	39.3	129	**118**	**12.5**
	18	29	13.8	123.1	31	6.5	123.1	31	19.4	123.1	**229**	**8.4**
	21	30	0.0	128.9	31	3.2	128.9	31	9.7	128.9	**222**	**6.2**
	n/aver.	n = **4**	aver. 12.6		n = **4**	aver. 12.2		n = **4**	aver. 27.3		n = **4**	aver. 12.6
Area IV	**22**				16	0.0	122.6	31	3.2	122.6	**47**	**1.6**
	23	30	0.0	144.2	14	0.0	144.2				**195**	**0.0**
	25	11	0.0	127.7							**162**	**1.1**
	n/aver.	n = **2**	aver. 0.0		n = **2**	aver. 0.0		n = **1**	aver. 3.2		n = **3**	aver. 0.9

In area I, the average percentage of porpoise-positive days per month was around 100% in September to November 2002, and in April to August 2003. It dropped to 70% in December 2002, to 39% in January 2003, and had its minimum of 23% in February 2003. In March 2003 the average percentage of porpoise-positive days per month rose to 52%.

In area II, the average percentage of porpoise-positive days per month was above 70% in August and September 2002; it declined to values below 10% for December 2002 to April 2003, and rose again above 60% until August 2003.

In area III, the average percentage of porpoise-positive days per month started with 62% in September 2002, dropped and stayed below 20% from November 2002 to July 2003, with minimum values for November 2002 and April/May 2003, and a rise in the winter months of 2002/2003.

In area IV, the average percentage of porpoise-positive days per month was near 0%, with one or two porpoise-positive days (resulting in up to 3%) in December 2002, as well as in January and August 2003.

4 Discussion

Our results show a clear decrease in the percentage of porpoise-positive days per month from the western part of the German Baltic Sea around Fehmarn to the eastern part up to the *Pommeranian Bay*, as well as seasonal changes around Fehmarn (area I) and in the *Kadet Trench* and adjacent coastal waters (area II). Verfuß et al. (2005) showed the importance of echolocation for harbour porpoises. Porpoises which were living in a well-known, semi-natural outdoor pool, permanently used echolocation even in easy orientation tasks during daylight regardless of the season. Therefore, a regular use of echolocation by harbour porpoises is likely. The changes in the amount of porpoise registrations in the course of the year and differences across areas are assumed to be caused by temporal changes and geographical differences in harbour porpoise density.

A decrease in harbour porpoise density from west to east in the German Baltic Sea is also confirmed by aerial surveys in 2002 and 2003 (see chapter 11). During the 2002 surveys, when Scheidat et al. (2004) observed aggregations of harbour porpoises in the *Pommeranian Bay*, no T-POD was deployed in this area. T-PODs deployed from November 2002 onwards showed only a few harbour porpoise registrations. Scheidat et al. (2004) registered no sighting in the *Pommeranian Bay* during their surveys after September 2002.

Morphological and genetic studies revealed the existence of a separate subpopulation of harbour porpoises in the Baltic proper, i.e., east of *Darss Sill* (*Darsser Schwelle*) (Huggenberger et al. 2002, Tiedemann et al. 2001). Low density of this subpopulation (see chapter 11) raises deep concern for the survival of the population, which is especially emphasised in the recovery plan for Baltic harbour porpoises (Jastarnia Plan, ASCOBANS). The T-POD data confirm a very low density of harbour porpoises in the German part of the Baltic proper. Any negative anthropogenic influence (e.g., incidental fishery by-catch, chemical or noise pollution) on this very small and therefore highly endangered subpopulation might sooner or later lead to its extinction.

Until the mid-20th century, migration of harbour porpoises was assumed for the North and Baltic Sea (reviewed in Koschinski 2003). In spring, the porpoises were thought to have followed movements of herring, passing Danish waters into the Baltic Sea. In late autumn and winter, when the Baltic tended to freeze over in some years, the porpoises may have migrated back out of the Baltic Sea. Nowadays, the porpoise stocks are too small to easily prove such migrations. Teilmann et al. (2004) could prove seasonality in the use of areas in Danish waters with the help of satellite tags on porpoises. Siebert et al. (in preparation) showed seasonality in incidental sightings and stranding rates in the German Baltic Sea, with a peak in the summer months. The data of incidental sightings might be biased by a lower effort in winter (e.g., less sailing boats), whereas stranding events can be biased by a longer submersion time of carcasses when water temperature is low (Moreno 1993, in Siebert et al. in preparation). The T-PODs proved seasonal changes in the use of the Baltic Sea areas around Fehmarn and the *Kadet Trench*.

The method of T-POD deployment proved to be a valuable tool for investigating the habitat use by harbour porpoises of the German Baltic Sea in a temporal and geographical scale. The results of this work revealed a regular use of the area around Fehmarn and the *Kadet Trench* for harbour porpoises in German waters. This showed the importance of these areas for these animals in Germany. A continuation of T-POD deployment, like the one presented in this study is necessary to confirm the revealed seasonal changes and geographical differences in harbour porpoise registrations, which is assumed to reflect differences in harbour porpoise densities. As a future goal, the inclusion of the *Kiel Bight* (*Kieler Bucht*) and the area around the island of Usedom is important for receiving a complete picture of the use of the German Baltic Sea by harbour porpoises. For investigating the highly endangered harbour porpoise subpopulation of the Baltic proper, the use of T-PODs has to be extended within the *Pommeranian Bay* by adding more T-POD measuring stations.

Acknowledgements

This work was funded by the German Federal Ministry for the Environment, Nature Conservation and Nuclear Safety (BMU), including the Investment-in-future program (ZIP) as part of the project MINOS, and by the Federal Agency for Nature Conservation (BfN). These data would not have been obtained without the administrative help of the Federal Maritime and Hydrographic Agency, the Coastguard Service, the Federal Border Guard, the Water and Shipping Authorities Stralsund and Lübeck. We are very grateful for the cooperation, assistance and discussions given by Nick Tregenza, and for the statistical help given by Roger Mundry. The following people were highly involved in data acquisition and analysis: Ines Baresel, Annette Kilian, Anja Meding and Peter Leopold. We thank Anita Gilles, Jonas Teilmann and an anonymous referee for valuable comments on the manuscript.

References

Au WWL (1993) The sonar of dolphins. Springer-Verlag, New York Berlin Heidelberg

Benke H, Siebert U (1994) Zur Situation der Kleinwale im Wattenmeer und in der südlichen Nordsee. In: Lozán JL, Rachor E, Reise K, von Westernhagen H, Lenz W (eds) Warnsignale aus dem Wattenmeer. Blackwell Wissenschaftsverlag, Berlin, pp 309–316

Benke H, Siebert U, Lick R, Bandomir B, Weiss R (1998) The current status of harbour porpoises (*Phocoena phocoena*) in German waters. Arch Fish Mar Res 46:97–123

Boye P, Hutterer R, Benke H (1998) Rote Liste der Säugetiere. Schriften-Reihe für Landschaftspflege und Naturschutz 55:33–39

Goodson AD, Kastelein RA, Sturtivant CR (1995) Source levels and echolocation signal characteristics of juvenile harbour porpoises (*Phocoena phocoena*) in a pool. In: Nachtigall PE, Lien J, Au WWL, Read AJ (eds) Harbour porpoises – laboratory studies to reduce bycatch, vol 1. De Spil Publishers, Woerden, The Netherlands, pp 41–53

Hammond PS, Berggren P, Benke H, Borchers DL, Collet A, Heide-Jorgensen MP, Heimlich S, Hiby AR, Leopold MF, Oien N (2002) Abundance of harbour porpoise and other cetaceans in the North Sea and adjacent waters. Journal of Applied Ecology 39:361–376

Heide-Jørgensen MP, Teilmann J, Benke H, Wulf J (1993) Abundance and distribution of harbour porpoises *Phocoena phocoena* in selected areas of the western Baltic and the North Sea. Helgoländer Meeresuntersuchungen 47:335–346

Huggenberger S, Benke H, Kinze CC (2002) Geographical variation in harbour porpoise (*Phocoena phocoena*) skulls: support for a separate non-migratory population in the Baltic proper. Ophelia 56:1–12

ICES/ACME (1997) Report of the ICES Advisory Committee on the marine environment. ICES Cooperative Research Report. ICES, Copenhagen. 222

Kamminga C, Engelsma FJ, Terry RP (1999) An adult-like sonar wave shape from a rehabilitated orphaned harbour porpoise (*Phocoena phocoena*). Ophelia 50:35–42

Kinze CC (1995) Exploitation of harbour porpoises (*Phocoena phocoena*) in Danish waters: a historical review. In: Bjorge A, Donovan GP (eds) Biology of the Phocoenids. Black Bear Press, Cambridge, pp 141–153

Koschinski S (2003) Current knowledge on harbour porpoises (Phocoena phocoena) in the Baltic Sea. Ophelia 55:167–197

Kröger R (1986) The decrease of harbour porpoise populations in the Baltic and North Sea. Final report on the Swedish-German harbour porpoise project funded by the World Wildlife Fund Sweden and Hamburg, University of Hamburg

Moreno P, Benke H, Lutter S (1993) Behaviour of harbour porpoise (*Phocoena phocoena*) carcasses in the German Bight: surfacing rate, decomposition and drift routes. In: Bohlken H, Benke H, Wulf J (eds) Untersuchungen über Bestand, Gesundheitszustand und Wanderung der Kleinwalpopulationen (*Cetacea*) in deutschen Gewässern. FKZ 10805017/11. BMU-Final Report. Institut für Haustierkunde, University of Kiel, Germany

Reijnders PJH (1992) Harbour porpoises *Phocoena phocoena* in the North Sea: numerical responses to changes in environmental conditions. Neth J Aquatic Ecol 26:75–85

Scheidat M, Gilles A, Siebert U (2004) Teilprojekt 2 – Erfassung der Dichte und Verteilungsmuster von Schweinswalen (*Phocoena phocoena*) in der deutschen Nord- und Ostsee. In: Kellermann A, et al. (eds) Marine Warmblüter in Nord- und Ostsee: Grundlagen zur Bewertung von Windkraftanlagen im Offshore-Bereich. FKZ 0327520. Final report. Investment-in-future program of the German Federal Ministry for the Environment, Nature Conservation and Nuclear Safety (BMU). Landesamt für den Nationalpark Schleswig-Holsteinisches Wattenmeer, Tönning. (www.wattenmeer-nationalpark.de)

Schulze G (1996) Die Schweinswale. Westarp-Wissenschaften, Magdeburg

Siebert U, Benke H, Schulze G, Sonntag RP (1996) Über den Zustand der Kleinwale. In: Lozán JL, Lampe R, Matthäus W, Rachor E, Rumohr H, von Westernhagen H (eds) Warnsignale aus der Ostsee. Parey Buchverlag, Berlin, pp 242–248

Siebert U, Gilles A, Lucke K, Ludwig M, Benke H, Kock K-H, Scheidat M (in preparation) Review of occurrence of the harbour porpoise (*Phocoena phocoena*) in German waters – analyses of aerial surveys, incidental sightings and strandings

Sonntag RP, Benke H, Hiby AR, Lick R, Adelung D (1999) Identification of the first harbour porpoise (*Phocoena phocoena*) calving ground in the North Sea. J Sea Res 1–8

Teilmann J, Dietz R, Larsen F, Desportes G, Geertsen BM, Andersen LW, Aastrup PJ, Hansen JR, Buholzer L (2004) Satellitsporing af marsvin i danske og tilstødende farvande. Danmarks Miljøundersøgelser. Report. DMU 484, Electronic version: http://www2.dmu.dk/1_viden/2_Publikationer/3_fagrapporter/abstrakter/abs_484_DK.asp

Tiedemann R (2001) Stock definition in continuously distributed species using molecular markers and spatial autocorrelation analysis. SC/53/SD3, London Paper presented to the Scientific Committee of the International Whaling Commission

Verfuß UK, Miller LA, Schnitzler H-U (2005) Spatial orientation in echolocating harbour porpoises (*Phocoena phocoena*). Journal of Experimental Biology 208(17):3385–3394

Verfuß UK, Schnitzler H-U (2002) R+D project: Untersuchungen zum Echoortungsverhalten der Schweinswale (*Phocoena phocoena*) als Grundlage für Schutzmaßnahmen. FKZ: 898 86 021, Final report. Eberhard Karls University of Tübingen, Dept. Animal Physiology, Germany. (Available at http://www.bfn.de/09/fkz89886021_end.pdf)

Verfuß UK, Honnef C, Benke H (2004a) R+D project: Untersuchungen zur Nutzung ausgewählter Gebiete der Deutschen und Polnischen Ostsee durch Schweinswale mit Hilfe akustischer Methoden. FKZ: 901 86 020. Final report, German Oceanographic Museum, Stralsund, Germany

Verfuß UK, Honnef C, Benke H (2004b) Teilprojekt 3 – Untersuchungen zur Raumnutzung durch Schweinswale in der Nord- und Ostsee mit Hilfe akustischer Methoden (PODs). In: Kellermann A, et al. (eds) Marine Warmblüter in Nord- und Ostsee: Grundlagen zur Bewertung von Windkraftanlagen im Offshore-Bereich. FKZ 0327520. Final report. Investment-in-future program of the German Federal Ministry for the Environment, Nature Conservation and Nuclear Safety (BMU). Landesamt für den Nationalpark Schleswig-Holsteinisches Wattenmeer, Tönning. (www.wattenmeer-nationalpark.de)

Chapter 13

Identification of areas of seabird concentrations in the German North Sea and Baltic Sea using aerial and ship-based surveys

Stefan Garthe

Research and Technology Centre Westcoast (FTZ), Büsum

Abstract

This paper gives a brief overview of the field methods used to study the distribution of seabirds at sea in the German parts of the North and Baltic Seas. It demonstrates how the data were analysed, how seabird concentrations may be delineated, and how suggestions for protected areas were derived from the data.

Seabird distribution was studied by transect counts from ships and aircraft. Species distribution maps produced from these data are based on densities. The distribution of widely dispersed species, e.g., lesser black-backed gull (*Larus fuscus*) and northern fulmar (*Fulmarus glacialis*), were analysed using grid maps. For species occurring in concentrations, a spatial interpolation procedure using ordinary kriging[1] was adopted. Examples of distributions and maps are given for long-tailed duck (*Clangula hyemalis*; Baltic Sea), common eider (*Somateria mollissima*; Baltic Sea) and red-throated and black-throated divers (*Gavia stellata* and *G. arctica*; North Sea). For all specially important species (i.e., species listed in Annex I of the EU Birds Directive that should be the subject of special conservation measures, e.g., red-throated diver and sandwich tern *Sterna paradisaea*), concentration areas were identified and subsequently combined so that a set of potential areas for conservation could be determined. From this map, potential Special Protection Areas (SPAs) were established.

Finally, this paper briefly discusses field methods and methods of analysis and gives further recommendations.

---

[1] Kriging: a form of statistical modelling that interpolates data from a known set of sample points to a continuous surface.

1 Introduction

Marine protected areas (MPAs) for seabirds are currently being established under various international instruments and marine conventions (e.g., OSPAR Convention, Helsinki Convention) and also under the main nature conservation directives of the European Commission. When Germany adopted its Federal Nature Conservation Act in April 2002 in order to select NATURA 2000 sites (including SPAs for birds) also within the Exclusive Economic Zone (EEZ), the need arose to get an up-to-date overview of the distribution and status of seabirds in the German North and Baltic Sea waters. This paper briefly describes the field methods used in studying the distribution of seabirds at sea, how the data were analysed, how seabird concentrations may be identified, and how suggestions for protected areas were derived from the data. The main emphasis is on methods rather than on results. The latter is presented in much more detail in Garthe (2003), Garthe et al. (2003), and Garthe and Skov (in preparation).

2 Materials and methods

2.1 Field methods

Seabird distribution in the southeastern North Sea and the southwestern Baltic Sea was studied by transect counts from ships and aircraft. These methods basically aim at assessing distribution patterns and numbers of seabirds at sea, but they are each differently suited for these purposes (see Camphuysen et al. 2004, and Garthe et al. 2004 for recent reviews). Aerial surveys are able to cover much larger areas in much shorter times at lower per-kilometre price. However, they are only feasible at low-wind situations and there are restrictions on species identification from aerial surveys (e.g., groups such as grebes, gulls, terns, and auks usually cannot be identified up to species level). Ship-based surveys enable collecting additional information on the behaviour of the birds and usually allow for sampling environmental data such as hydrography, which proves very useful for understanding species distribution patterns.

For counts from ships, the methodology has been largely standardised internationally and was first described by Tasker et al. (1984). Due to the presence of high densities of birds that quite often fled away from approaching ships, it proved necessary to regularly or continuously search for birds using binoculars and to deploy at least two observers, as suggested by Webb and Durinck (1992) and Garthe et al. (2002). Birds were counted from either the top deck or the bridge-wing, usually on 300-metre wide transects set to one or both sides of the vessel. Counting intervals during the surveys were initially set at 10 minutes, as suggested by Tasker et al. (1984), but 1- or 2-minute intervals were increasingly used to enable higher resolution for mapping and analysis of seabird distribution, particularly in relation to water depth and hydrography. Flying birds were counted employing the *snapshot* method (Tasker et al. 1984, Garthe et al. 2002). Positions of the survey vessel were recorded automatically by onboard or portable GPS instruments.

Seabirds were also counted from aircraft using a transect methodology recently described by Diederichs et al. (2002). Flights were conducted from double-engine planes (e.g., Partenavia P-68) flying over German waters, from the coast to the outer limit of the EEZ. Transects were usually set perpendicular to the coast to obtain variation over major habitat features such as water depth, distance to coast, and frontal systems. Transects were separated in the North Sea by 10 km (20 km in areas far from the coast) and in the Baltic Sea mostly by 8 km. Flights were conducted at an altitude of 250 feet (78 metres) and a speed of 100 knots (185 km/hour). Transect bands were set for each observer by inclinometers (devices measuring angles) on the side of the aircraft since there was no possibility of looking at the sea surface directly under the plane. Transect widths were either 122 metres (one band only) or 397 metres (two bands) depending on viewing conditions. During the counts, all bird observations were recorded on a portable voice recorder; data recorded include: time (to the nearest second), species, number, general behaviour (five categories) and also, if possible, age and sex. Geographic position was recorded every 5 seconds on-board the plane.

2.2 Data bases

The data bases for the work described in this paper are current versions of the European Seabirds-at-Sea (ESAS) Database, the German Seabirds-at-Sea Database, and the BALTSAS-Database (DHI Water & Environment, Hørsholm, Denmark, c/o H. Skov) (see e.g., Durinck et al. 1994, Stone et al. 1995, Garthe and Hüppop 2000). In Germany, most data were collected by the Research and Technology Centre (FTZ) in Büsum (an external station of the University of Kiel) and the Institute of Avian Research; in Denmark, by Ornis Consult Ltd; in the UK, by the Joint Nature Conservation Committee (JNCC); and in the Netherlands, by the Royal Netherlands Institute for Sea Research (RNIOZ), Camphuysen Seabird Research (CSR), and ALTERRA.

In total, the following distances were covered: in the southeastern North Sea, 103,000 km by ship from 1990 to June 2004, and 28,000 km by plane from 2002 to June 2004; in the southwestern Baltic Sea, 19,000 km each by ship from 1986 to June 2004, and by plane from 2002 to June 2004. The effort at sea varied substantially over the years and seasons.

2.3 Selection of species for SPA delineation

Gellermann et al. (2003) catalogued those species whose occurrence in German waters needed to be considered for the selection of SPAs. They distinguished three different levels of species importance. For the selection of SPAs, only those species that were categorised in their list as of high or medium importance were used. Three groups of bird species comprised this category.

Firstly, the species that are currently listed in Annex I of the EU Birds Directive (species that shall be the subject of special conservation measures) and that occur regularly in the offshore waters of the German parts of the Baltic Sea. These are the red-throated diver (*Gavia stellata*), black-throated diver (*Gavia arctica*), slavonian grebe (*Podiceps auritus*), and four species of terns. (Please note that grebes and terns are much more restricted to the coast in the North Sea than they are in the Baltic Sea).

Secondly, migratory species which regularly occur in offshore areas were included. The EU Birds Directive does not define "migratory species", and the definition used in practice is the one provided by the Bonn Convention. This convention defines migratory species as species where a significant proportion of the population cyclically and predictably crosses one or more national jurisdictional boundaries. For both study areas, this includes all seabird species. Then, for these bird species, especially those that occur in major concentrations, the most important areas (or a few of the most important areas) were recommended to be selected as SPAs. The identification of SPAs focused on the German EEZ, and preferably, on areas derived from more than one species.

The third group or category refers to rare offshore species and species occurring along the coast only (e.g., diving ducks, geese, swans). Analysis of data was equal for the first two groups, i.e., the Annex I-bird species and migratory species. However, concentrations of Annex I-bird species were considered to be much more important and were thus more decisive for the designation of SPAs than areas with only migratory bird species. The third group (i.e., rare offshore and along-the-coast-only species) was not at all relevant for the SPA selection process in the EEZ due to their virtual absence in that area.

2.4 Species distribution maps

All species distribution maps are based on densities, that is, the number of individuals per unit area. Principally, two types of seabird distribution patterns may be distinguished. Some species are distributed over large areas and exhibit usually only short-term aggregations (e.g., gulls), while other species are often heavily concentrated and are reliably predictable in their distribution (e.g., sea ducks). Although there are intermediate patterns, different approaches were adapted to map and they delineate the major distribution areas. For widely distributed species, distributions were analysed by grid maps with grid cells by either 3′ latitude x 5′ longitude (grid cell size: ca. 30 km^2) or 6′ latitude x 10′ longitude (grid cell size: ca. 120 km^2). Species with wide-ranging distributions were visualised by the larger grid cells, and species with a more restricted distribution by the smaller grid cells. For each grid cell, the overall density was calculated, obtained from the sum of all birds recorded in transect divided by the total area mapped. This way, the data was corrected for effort. Major areas for widely distributed species were difficult to determine precisely.

All species that occur in concentrations aggregated in areas that were far too large to allow total counts of all the birds. For these species, therefore, a spatial interpolation procedure based on ordinary kriging (Kitadinidis 1997, and used by Skov et al. 2000) was adopted and further developed (Garthe 2003, Garthe and Skov in preparation). With this procedure, distributional data were interpolated and smoothed between survey lines on the basis of the species-specific spatial abundance structure (which is measured by the software used).

2.5 Seabird concentrations

Boundaries of concentration areas were determined by an analysis investigating the gradient of density change over space. In order to do that, the modelled distributional data were projected into a two-dimensional map. In each of such cases, the modelled isoline of bird density (i.e., the line drawn through the same level of bird density) located just outside the strongest gradient in spatial density was chosen as the border of a concentration. In this way, the major part of the concentration is included in the selected area. The density value of the borderline was noted and used as the species- and season-specific minimum density defining a seabird concentration. This value was then taken for plotting the contour line showing the spatial extent of the respective concentration.

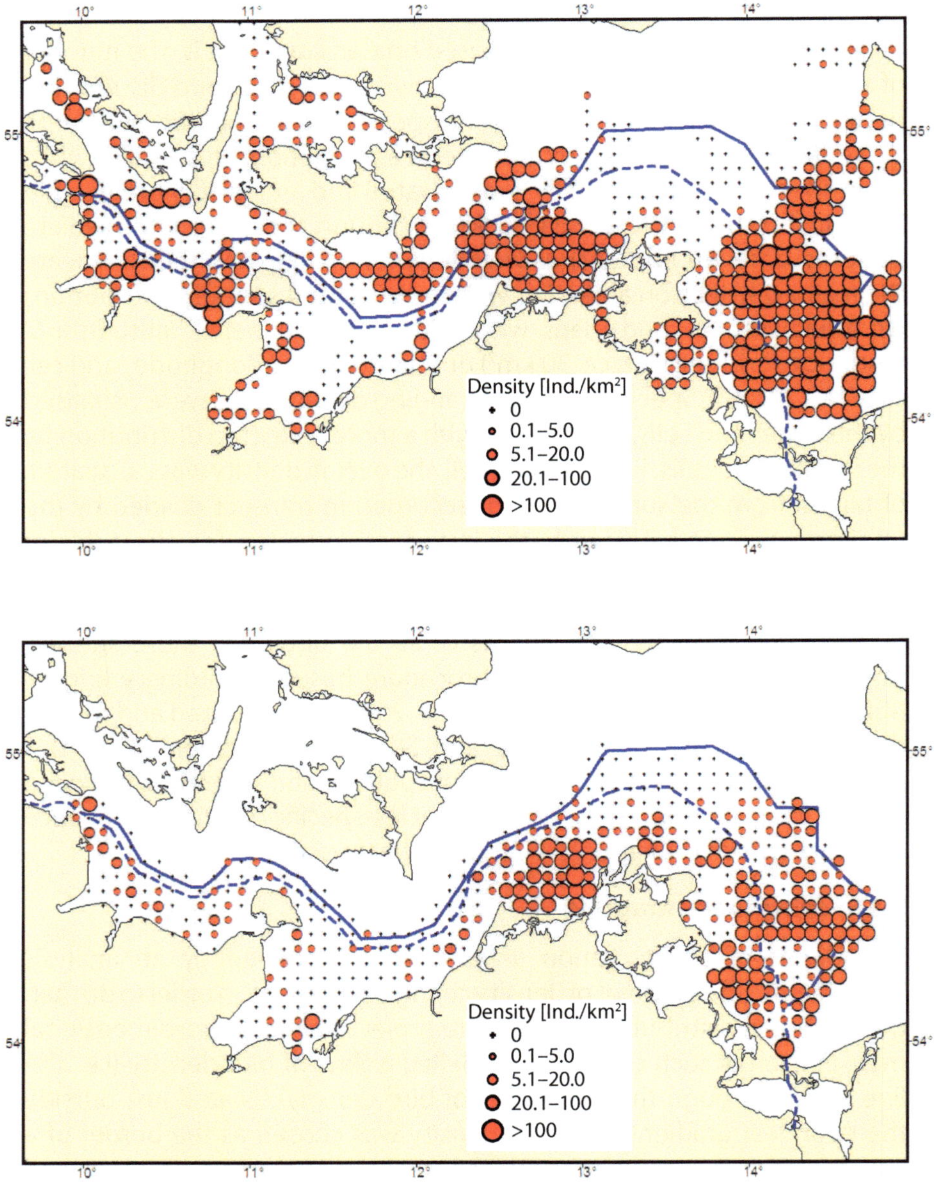

Figure 1. Distribution of long-tailed ducks in the southwestern Baltic Sea in winter as assessed by two different methods. The dashed line indicates the 12-mile zone, the continuous line the EEZ border. (Top) Ship-based data for November–February 1987–2003, (bottom) aerial survey data for February 2003

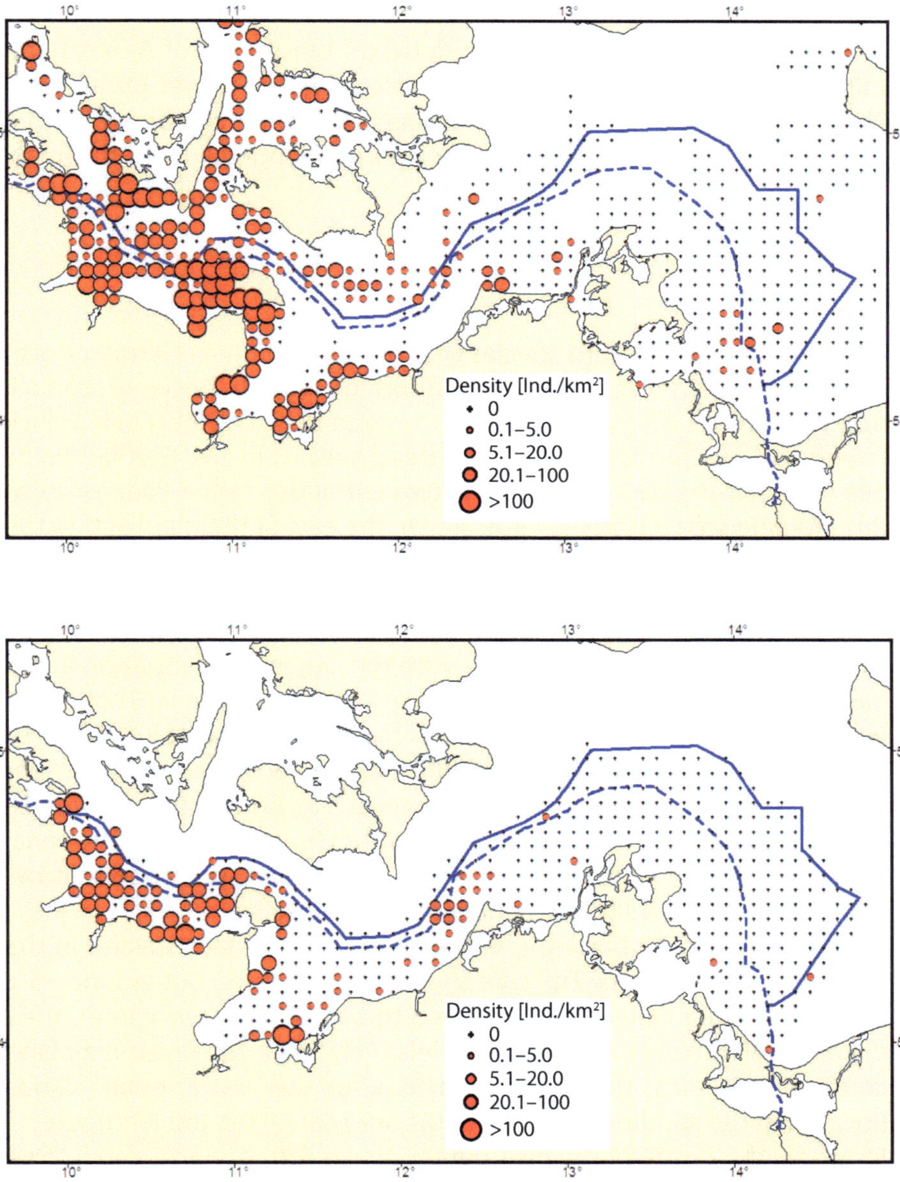

Figure 2. Distribution of common eiders in the southwestern Baltic Sea in winter as assessed by two different methods. The dashed line indicates the 12-mile zone, the continuous line the EEZ border. (Top) Ship-based data for November–February 1987–2003, (bottom) aerial survey data for February 2003

2.6 Combining single-species data into multi-species data

For each of the species of interest derived from the list of Annex I and migratory bird species, concentration areas were retained for analysis. These respective areas and contour lines were then combined so that a set of areas for potential conservation was identified. From this map, potential SPAs were derived.

3 Results

From a total of 25 seabird species that occur regularly in German North and Baltic Sea offshore waters, the distributions of two species of seaduck and two species of divers are shown as examples. The long-tailed duck (*Clangula hyemalis*) is a fairly widespread species in the German Baltic Sea in winter (figure 1), with major concentrations in the *Pommeranian Bay* (*Pommersche Bucht*; the large area in the east of the map) and to the west of Rügen (located towards the centre of the map). There are a few more but less obvious concentration areas. In contrast, the common eider (*Somateria mollissima*) is restricted in its distribution to the western part of the German Baltic Sea (figure 2), with the largest concentration in the *Kiel Bight* (*Kieler Bucht*) (southwestern part of the study area).There were very few sightings of this species east of the Darss peninsula in winter. For both species, the results are quite similar as derived from a multiple year data set covering waters to different extents (top figures 1 and 2) as well as by a single three-day aerial survey (bottom figures 1 and 2). The aerial survey took place at a time when there was no ship survey so that the data sets are temporally independent in that sense.

Red-throated and black-throated divers are important species in the southeastern North Sea. They are shown in joint maps, but around 95% of the individuals identified were red-throated divers. In winter, their distribution is mainly restricted to a relatively small zone near mainland coasts and islands; this is most obvious for the East Friesian Islands (located in the southeastern part of the study area; see top left figure 3). In spring, the centre of the distribution moves further north and further away from the coast and is located west of the North Friesian Islands in the EEZ (top right and bottom figure 3).

Figure 4 shows the concentration areas of all relevant seabird species in the southwestern Baltic Sea following the methodology described in sections 2.3, 2.5 and 2.6. These species are red-throated diver, black-throated diver, great crested grebe (*Podiceps cristatus*), red-necked grebe (*Podiceps grisegena*), slavonian grebe, common eider, long-tailed

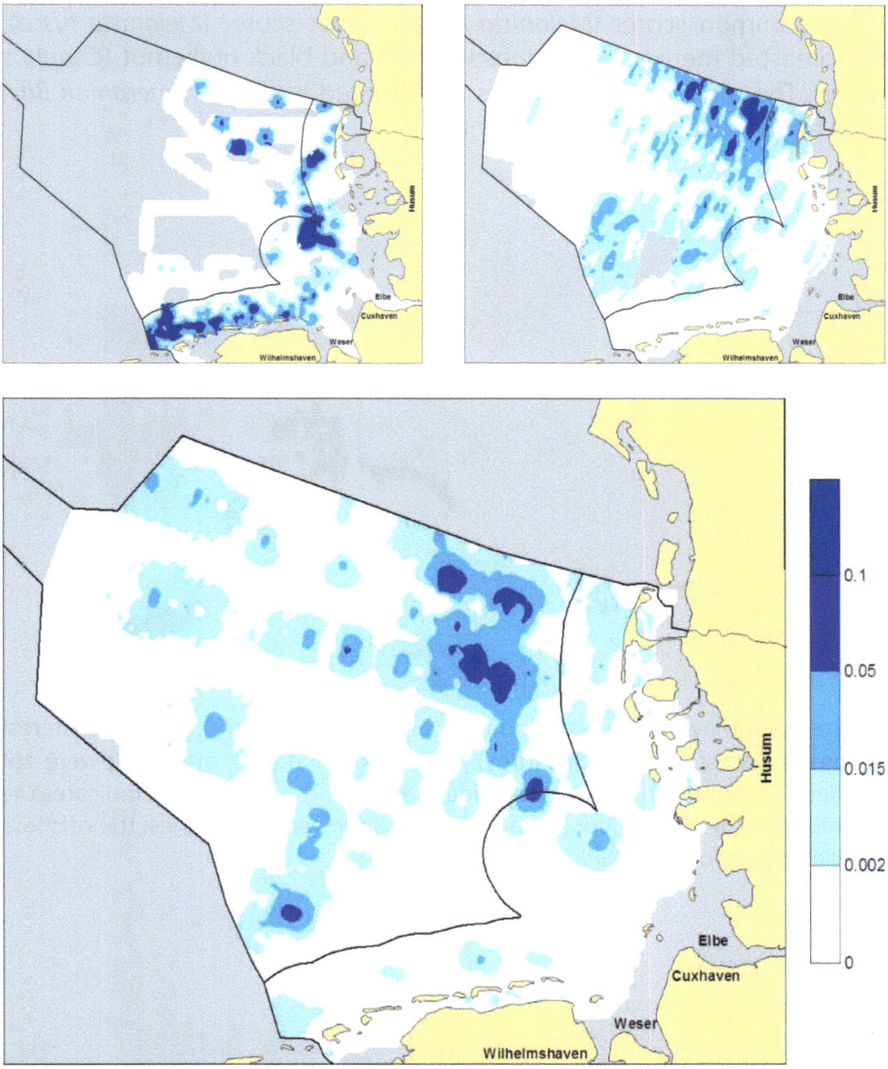

Figure 3. Distribution of red-throated and black-throated divers in the south-eastern North Sea. Data originated from two different methods. The maps have been produced by spatial interpolation; for details see text. Light grey colour indicates areas that were not studied sufficiently during the respective survey periods and/or were outside of the scope (e.g., areas in countries other than Germany). The line more or less parallel to the coast indicates the 12-mile zone, the other line the EEZ border. The *x*-axis gives a total distance of 290 km, the *y*-axis 245 km. The colour scale gives the abundance of the birds on a logarithmic scale (log (birds*km^{-2}+1). (Top left) November–February 2000–2003, ship data; (top right) March 2003, aerial survey data; (bottom) April 2003, aerial survey data

duck, common scoter (*Melanitta nigra*), velvet scoter (*Melanitta fusca*), red-breasted merganser (*Mergus serrator*) and black guillemot (*Cepphus grylle*). The most important areas are situated in the *Pommeranian Bay*, west of Rügen, and in the *Kiel Bight*.

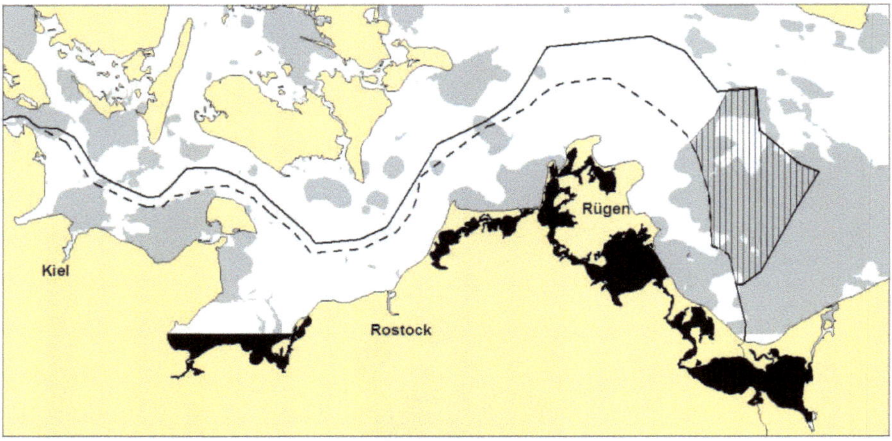

Figure 4. Overlay of all areas exhibiting concentrations of the species of interest in the southwestern Baltic Sea (grey). Black colour indicates areas that were not studied sufficiently and/or were outside the scope. The dashed line indicates the 12-mile zone, the continuous line the EEZ border. Vertical hatches: the offshore Special Protection Area (SPA) for birds

4 Discussion

The data collected by ship-based and aerial surveys have been extremely useful for describing the current distribution patterns of all seabird species in the two study areas. Both field methods have their advantages and disadvantages because of their different characteristics (see section 2.1). A combination of both is ideal for most purposes, including the ones discussed in this paper. It is important to carefully select these field methods with respect to the spatio-temporal scale envisaged for such a study, and to the species under consideration.

The robustness of the results has been the focus of considerable attention by different groups in light of proposals for two large SPAs within the German EEZs of the Baltic and North Seas (Garthe 2003). Most promising was the finding that all surveys carried out after finalisation of the SPA proposals (i.e., all surveys in 2003 and 2004) could prove the same major concentration areas as in the years before (e.g., top right and bottom figure 3 for the North Sea). This shows that even if borderlines of concentration areas may shift slightly, as to be expected by seabirds living in a dynamic environment, the major results are stable and reproducible. However, on a larger time-scale, especially if environmental conditions change, it is possible that the distribution of seabird species may alter. In the Baltic Sea, this could be the case, for instance, with regard to winter ice distribution since nearly all data collected for this study were from mild and normal winters only. In the North Sea, recent evidence of major changes in food availability (which have led to breeding failures in the northwestern North Sea) may influence distribution patterns in the *German Bight* (*Deutscher Bucht*), too, at least of those species ranging over wide areas of the North Sea (northern fulmar, northern gannet *Morus bassanus*).

The analytical methods outlined in this paper are still at an early stage of being adopted as standard procedures for designating SPAs since most Member States of the European Union have not yet delineated such protected sites in offshore areas. However, these methods have been very useful for selecting concentration areas of seabirds and SPAs. More recent works by British colleagues highlight the way forward (e.g., McSorley et al. 2004, Webb et al. in preparation). For species exhibiting widely dispersed distributions, the procedure for identifying concentration areas is much more difficult than for aggregated species. Up to now, no attempt has been shown on how to deal with sea areas where only species that are rather evenly or at least widely distributed occur. In such cases, vast areas would need to be designated to capture a given percentage of bird numbers. This is often politically impossible and might also be scientifically less evident. This problem needs further consideration. For modelling purposes, e.g., future site selections, co-variates (e.g., water depth) should be included. Also, attention may be paid to the reliability of the data by calculating (statistically) the spatial variation of the boundary lines describing concentrations. If such boundary lines vary substantially over space within e.g., standard deviation, then the data basis and/or the aggregation character of the bird species may be less evident than when the boundaries are more stable over space.

Figure 5. Northern Gannet (*Morus bassanus* (L))

Acknowledgements

This work is based on several projects that have been conducted in recent years. Funding has been received primarily by Bundesamt für Naturschutz, Bundesministerium für Umwelt, Naturschutz und Reaktorsicherheit, Freunde und Förderer der Inselstation der Vogelwarte Helgoland e.V., Ornithologische Arbeitsgemeinschaft für Schleswig-Holstein & Hamburg e.V. and Forschungs- und Technologiezentrum Westküste (FTZ) der Universität Kiel. Field observations were carried out by many observers over the years. Many private and governmental institutions permitted work on their boats. Furthermore, many people have contributed to this work by collecting, summarising and analysing data, by helping shape the procedures for SPA designations and through general comments. From these, at least the following need to be mentioned: C.J. Camphuysen, V. Dierschke, O. Engelhard, N. Guse, O. Hüppop, J. Kotzerka, J. Krause, U. Kubetzki, K. Ludynia, N. Markones, T. Merck, M. Scheidat, P. Schwemmer, H. Skov, N. Sonntag, A. Webb, and T. Weichler.

References

Camphuysen CJ, Fox AD, Leopold MF, Petersen IK (2004) Towards standardised seabirds at sea techniques in connection with environmental impact assessments for offshore wind farms in the U.K. Final report to COWRIE, COWRIE-BAM-02-2002

Diederichs A, Nehls G, Petersen IK (2002) Flugzeugzählungen zur großflächigen Erfassung von Seevögeln und marinen Säugern als Grundlage für Umweltverträglichkeitsstudien im Offshorebereich. Seevögel 23:38–46

Durinck J, Skov H, Jensen FP, Pihl S (1994) Important marine areas for wintering birds in the Baltic Sea. Ornis Consult report, Kopenhagen

Garthe S (2003) Verteilungsmuster und Bestände von Seevögeln in der Ausschließ-lichen Wirtschaftszone (AWZ) der deutschen Nord- und Ostsee und Fach-vorschläge für EU-Vogelschutzgebiete. Berichte zum Vogelschutz 40:15–56

Garthe S, Hüppop O (2000) Aktuelle Entwicklungen beim "Seabirds-at-Sea"-Programm in Deutschland. Vogelwelt 121:301-305

Garthe S, Skov H (in preparation) Selection of suitable sites for marine protected areas for seabirds: a case study of an EU Special Protection Area in the Baltic Sea

Garthe S, Hüppop O, Weichler T (2002) Anleitung zur Erfassung von Seevögeln auf See von Schiffen. Seevögel 23:47–55

Garthe S, Ullrich N, Weichler T, Dierschke V, Kubetzki U, Kotzerka J, Krüger T, Sonntag N, Helbig AJ (2003) See- und Wasservögel der deutschen Ostsee – Verbreitung, Gefährdung und Schutz. Federal Agency for Nature Conservation, Bonn

Garthe S, Dierschke V, Weichler T, Schwemmer P (2004) Rastvogelvorkommen und Offshore-Windkraftnutzung: Analyse des Konfliktpotenzials für die deutsche Nord- und Ostsee. Unpublished final report of sub-project 5 of the project MINOS to the Federal Environmental Ministry, Berlin

Gellermann M, Melter J, Schreiber M (2003) Ableitung fachlicher Kriterien für die Identifizierung und Abgrenzung von marinen Besonderen Schutzgebieten (BSG) nach Art. 4 Abs. 1 und 2 der Vogelschutzrichtlinie bzw. Vorschlagsgebieten gemäß Art. 4 Abs. 1 der FFH-Richtlinie für die deutsche ausschließliche Wirtschaftszone. Unpublished Report to the Federal Environ-mental Ministry, Berlin

Kitanidis P, (1997) Introduction to geostatistics – Applications in hydrogeology. Cambridge University Press, New York

McSorley CA, Webb A, Dean BJ, Reid JB (2004) Inshore marine Special Protection Areas: a methodological evaluation of site selection and boundary determination. JNCC Report 344. Joint Nature Conservation Committee, Peterborough

Skov H, Vaitkus G, Flensted KN, Grishanov G, Kalamees A, Kondratyev A, Leivo M, Luigojoe L, Mayr C, Rasmussen JF, Raudonikis L, Scheller W, Sidlo PO, Stipniece A, Struwe-Juhl B, Welander B (2000) Inventory of coastal and marine important bird areas in the Baltic Sea. BirdLife International, Cambridge

Stone CJ, Webb A, Barton C, Ratcliffe N, Reed TC, Tasker ML, Camphuysen CJ, Pienkowski MW (1995) An atlas of seabird distribution in north-west European waters. Joint Nature Conservation Committee, Peterborough

Tasker ML, Jones PH, Dixon TJ, Blake BF (1984) Counting seabirds at sea from ships: a review of methods employed and a suggestion for a standardized approach. Auk 101:567–577

Webb A, Durinck J (1992) Counting birds from ships. In: Komdeur J, Bertelsen J, Cracknell G (eds) Manual for aeroplane and ship surveys of waterfowl and seabirds. IWRB Special Publication, Slimbridge, UK 19:24–37

Webb A, McSorley CA, Dean BJ, Reid JB, Cranswick PA, Smith L, Hall C (in preparation) An assessment of the numbers and distributions of inshore aggregations of waterbirds using Liverpool Bay during the non-breeding season in support of possible SPA identification. JNCC Report. Joint Nature Conservation Committee, Peterborough

Chapter 14

The MINOS project: ecological assessments of possible impacts of offshore wind energy projects

Adolf Kellermann[a], Kai Eskildsen[b] and Barbara Frank[b]

[a] *International Council for the Exploration of the Sea (ICES), Copenhagen, Denmark*
[b] *Landesamt für den Nationalpark Schleswig-Holsteinisches Wattenmeer, Tönning, Germany*

Abstract

The joint research project MINOS examines whether large-scale offshore wind farms within the German parts of the North and Baltic Seas affect or endanger harbour porpoises, common seals, and seabirds. The research results are expected to provide the basis for estimating and assessing the impacts of future wind farms. MINOS focuses on two items: recording the preferential habitats and migratory routes of these animals in the EEZ, and investigating the sensitivity of porpoises and seals to sound, in order to assess possible damage, displacement and disturbance. When evaluating overall impacts to these animals, any expected impact of offshore wind farms must be considered in the context of already existing stressors.

The goal of the research is not to prevent or to hamper wind farming, but to provide a profound and reproducible knowledge basis for assessment in order to facilitate the development of sustainable power generation. This decision will not be taken by biologists or geologists, but by authorities or in court.

Since harbour porpoises have an ultrasonic location system (like bats), they are very sensitive to underwater noise. The noise produced during the construction and operation of offshore wind turbines, could cause behavioural changes or even physical harm to these animals. Such disturbances could displace the porpoises from their feeding and breeding habitats or otherwise reduce their fitness due to higher stress. Similar effects could also be expected to apply to common seals.

Preliminary results indicate that harbour porpoises and common seals avoid sources that emit sounds similar to that of wind turbines.

With regard to seabirds, MINOS has focused on divers and sea ducks, which have important wintering grounds in the North and Baltic Seas within the German EEZ. The construction of offshore wind farms within these areas should be assessed with regard to possible detrimental consequences caused by the loss or interruption of diver and sea duck resting and feeding habitats.

MINOS has also developed and expanded upon research and evaluation methods for future use in monitoring programmes. Resting seabirds and harbour porpoises were counted using low altitude aerial transect surveys. Complex mathematical modelling was used to estimate the abundance of animals. Similarly, ship surveys were also carried out. Telemetry was used to record spatial activity patterns of seals. Hearing tests were carried out with free-ranging porpoises and seals and their conspecifics in captivity. Finally, porpoise detectors (POD) were employed to detect the presence of porpoises in the vicinity. These detectors can function for several weeks, but cannot determine the relative number of animals; nonetheless, they provide important additional results to the numbers gained through airborne and at-sea surveys.

The above-described research has been conducted in seven separate subprojects, integrated into the MINOS-project and its successor, MINOSplus (see chapters 10, 11, 12 and 13). All projects were funded by the German Federal Ministry for the Environment, Nature Conservation and Nuclear Safety.

1 Introduction

The gradual departure from reliance on nuclear power is inevitably linked to exploiting renewable energy, if an additional goal is long-term climate protection. Using wind energy is very promising but not without side effects. Thus, research is needed to ensure that the development of wind energy projects complies with nature and environmental standards. Wind is plentiful in the north German coastal States and offshore wind energy has been viewed as an attractive prospect for a different reason: since most constant and unobstructed wind blows across the open sea, using wind energy offshore is most efficient, more so since the capacity of nature and landscape on this side of the dikes seems virtually exhausted. The coastal waters of the North and Baltic Seas enjoy certain conservation protection because of their beauty and uniqueness. Hence building wind farms in many of these areas is precluded, leading planners to consider

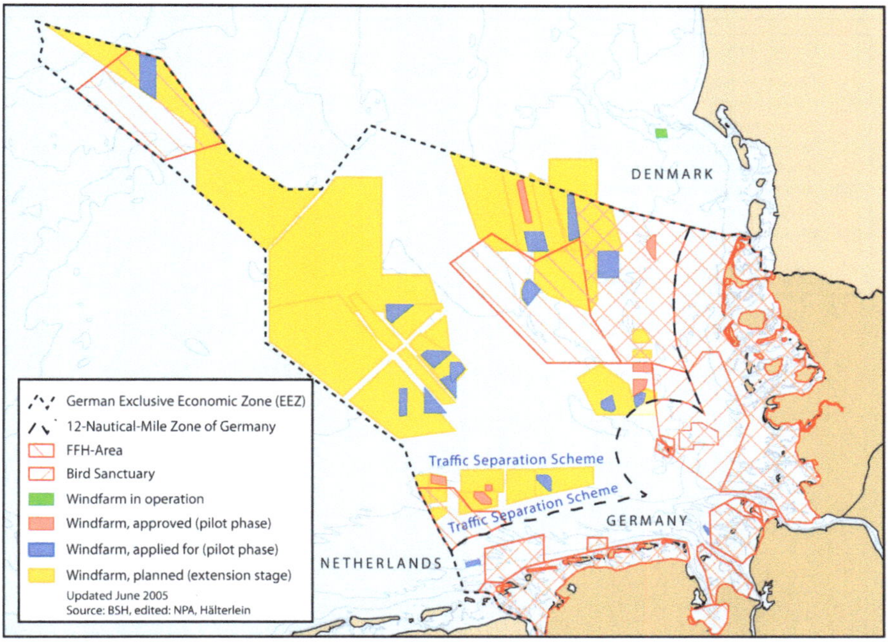

Figure 1. Status of offshore wind farming in the German EEZ of the North Sea in June 2005

mainly the Exclusive Economic Zone (EEZ) beyond the 12-nautical mile limit.

The EEZ, especially in the North Sea, already faces a wide array of utilisation and exploitation. For decades, there have been industrial activities such as oil and gas drilling and transport, cables and electrical fields, shipping lanes and fisheries which are sources of noise, mechanical impact, and pollution. Moreover, fisheries, for instance, reduce food resources of fish and marine mammals; and, shipping displaces flightless sea ducks. The construction, operation, and maintenance of offshore wind farms, as well as extensive cable nets only impose additional impacts on top of existing disturbances. The assessment of about 30 offshore wind farm applications, amounting to 6,000–7,000 turbines (figures 1 and 2), needs to take this background into account.

2 Background: the assessment of offshore wind farms

By 2001, it had become obvious that in order to properly assess their actual and potential effects on the marine environment, further

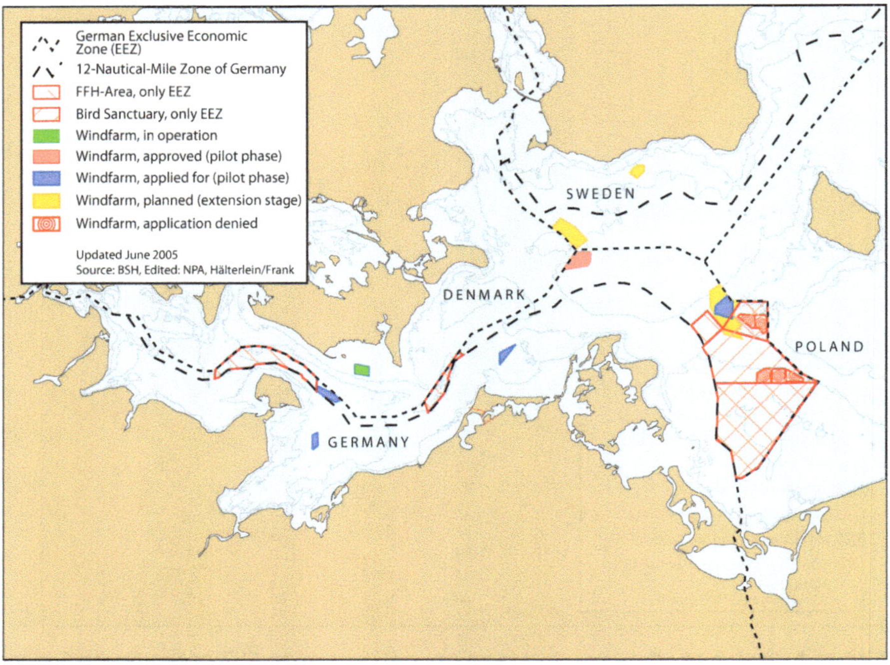

Figure 2. Status of offshore wind farming in the German EEZ of the Baltic Sea in June 2005

extensive research were required prior to the planning and construction of wind farms. Wind energy firms were obliged to conduct their own environmental impact assessment studies in areas adjacent to their claims (i.e., the future wind farms) on the basis of the so-called *standard investigation concept* (BSH 2003). As a result, our knowledge of localised marine seafloor fauna and of local distributions of mammals and seabirds is improving. However, because of the wide distribution and mobility of the species concerned (i.e., harbour porpoises, common seals, and seabirds), localised site-specific studies can render only a partial picture, making population-level assessments difficult or even impossible. Therefore, broad scale studies of multiple time series were (and continue to be) required for these species.

Rather optimistic scenarios have already been published, whereby in the future the rocky shore-like foundations of windmills host a rich flora and fauna that form the basis of an abundant food web, an effect enhanced by the lack of immediate fisheries. In such scenarios, apparently, seals and porpoises would prey on the plentiful fish, regardless of the noise and vibration caused by the turbines. This remains subject to speculation unless scientifically reliable data have been presented. Actually, a different scenario emerges. Installing the construction sites requires shipping and helicopter activities, ramming foundations and erecting the piles, etc., all of which will be accompanied by disturbances. Routes will be marked-out for cables, and networks of hundreds of miles will generate electrical fields with corresponding localised temperature effects in the sediment. During operation, each site, equipped with 100 or so windmills, will also require periodic maintenance. There will be noise and disturbance in abundance during construction, and vibrations and disturbance during operation, which currently can only be estimated theoretically.

Technologically, wind farming far offshore is a new challenge. While this provides an opportunity for the future export of such a technology, it must also be environmentally sustainable if it is to be truly competitive. Applying the experience from Scandinavian wind farms is of limited use due to conditions that are different from those in German waters. The wind farms in Denmark off Horns Rev and Nysted, for instance, are located closer to the coast and in shallower waters, while the German installations are planned in waters of around 20 metres in the open areas of the EEZ.

In early 2002, a total of seven research endeavours were launched and financed by the Federal Ministry of the Environment, Nature Conservation and Nuclear Energy which dealt with the ecological, technical and formal aspects of offshore wind farming. Two of these projects focused on the marine flora and fauna (Kutscher and Stump 2004). This chapter outlines one of them: the project MINOS, which is part of the German Federal government's Program for Future Investment (ZIP).

3 MINOS: marine mammals and birds in the North and Baltic Seas

In 2001, the National Park Regional Office for the Schleswig-Holstein Wadden Sea initiated investigations on harbour porpoises (*Phocoena phocoena*) in a newly established whale sanctuary off the islands of Sylt and Amrum. The sanctuary comprises 1,240 square kilometres of coastal waters where mother-calf groups were known to be numerous in spring and early summer. Acoustic devices were given a trial to record the frequency and patterns of calls. The first airborne survey had been conducted in the mid-nineteen nineties, and common seals have been the target of the Trilateral Monitoring and Assessment Program in the Wadden Sea since 1994 (TMAP, Kellermann 1995). Further, the Regional Office had begun to monitor the seabirds in the ebb deltas of the

National Park. Building on this previous research, the MINOS initiative was undertaken by the Regional Office in cooperation with its partners to conduct the study, "Marine Mammals and Birds in the North and Baltic Seas", as a basis for the assessment of offshore wind farms (figure 3).

MINOS has three objectives:

- To record the occurrence and behaviour of harbour porpoises and common seals in space and time in the North and Baltic Seas.
- To investigate the hearing-sensitivity and threshold of harbour porpoises and common seals, and the impact of acoustic emissions.
- To record the stocks and distribution of resting seabirds in space and time in the North and Baltic Seas.

MINOS was completed successfully in March 2004 (National Park Regional Office, 2004). A three-year continuation, called MINOSplus, started in June 2004 in order to fill gaps in knowledge, answer newly emerged questions, and prepare studies related to the construction phase of individual windmills. Like MINOS, MINOSplus consists of seven single projects that together form the joint project team (see chapters 10 to 13). The German Ministry for the Environment, Nature Conservation and Nuclear Safety funded both projects.[1]

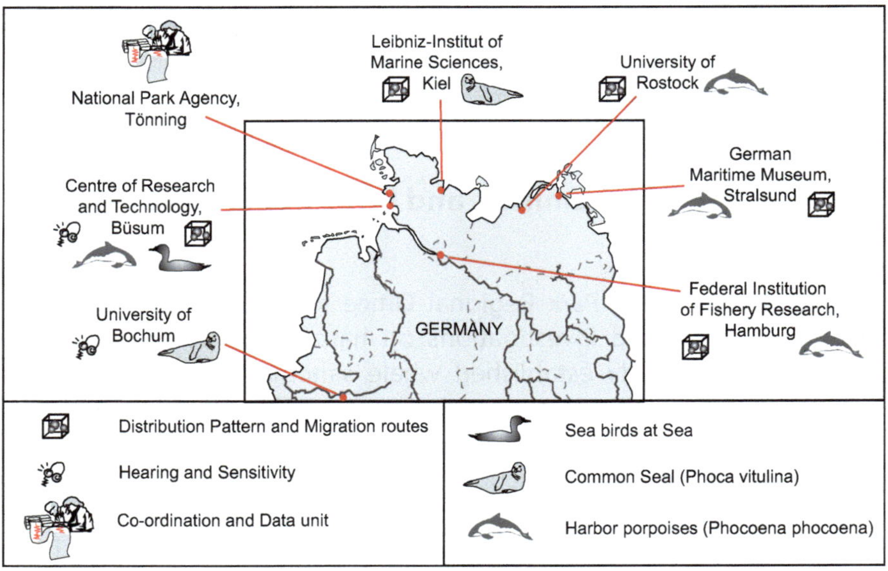

Figure 3. Overview of the institutions and research projects involved in MINOS

[1] Project funding numbers 0327520, 0329946, 0329946B-D.

4 Harbour porpoises

Harbour porpoises (*Phocoena phocoena*) are among the smallest toothed whales and are related to dolphins and sperm whales. They are distributed throughout the temperate seas of the northern hemisphere. In the Baltic Sea there may be two populations: one that extends approximately to the Darss barrier; and another eastern population in the Baltic proper that extends as far north as the Gulf of Finland. Presumably, harbour porpoises were once much more abundant. It is known that up to a century ago, they were herded and slaughtered in the bays of the Danish belt seas, (the same is still occurring with pilot whales in some islands even today). Nowadays, the main threat to harbour porpoises is the on-going by-catch in drift- and gill nets. They are seldom sighted, considered shy and presumably sensitive to disturbances. Pollutants and overfishing add to these stressors. Because harbour porpoises primarily orient themselves, communicate, and locate their prey using acoustic signals, impairment of their acoustic perception will have severe or even fatal consequences for individual porpoises and ultimately for the entire stock. Investigations have so far indicated that any further impacts contributing to increased mortality, reduced reproduction, or loss of habitat have to be prevented. Hence, the studies within MINOS are indispensable for minimising or avoiding such negative effects (Scheidat and Siebert 2003, also chapter 11).

Before MINOS began, information on the distribution and abundance of harbour porpoises in German waters was very limited. Field surveys included only few areas in the summer. The coastal waters off the islands of Sylt and Amrum were known for their regular occurrence of mother-calf sightings. For the first time, MINOS covered the entire EEZ using aerial surveys and acoustic methods. In contrast to previous assumptions that the porpoises were evenly spread across the North Sea, increased densities were observed in certain areas, especially in spring, which confirmed the importance of the Sylt and Amrum waters as a habitat of stable and regular occurrence. In the Baltic, acoustic surveys revealed the importance of certain hitherto unrecognised areas for harbour porpoises. Furthermore, there is a decline of abundance from west to east and the status and survival of the eastern population in the Baltic proper is uncertain (chapter 12). Further understanding of the large-scale variability and migration patterns requires additional research, such as provided by the continuation of the project. The importance of observing harbour porpoise distribution at broad scales has been demonstrated by MINOS as well as by the studies at the Danish wind farm of Horns Rev, where effects of construction works have not only been found near the site but also in distant reference areas (Tougaard et al. 2003).

The hearing sensitivity of harbour porpoises, which is recorded experimentally, is of particular importance for impact assessments, given the high level of noise of certain construction methods and the background noise during operation. Hearing sensitivity experiments have proven to be rather complex, requiring long periods of accommodation and training. Moreover, the number of free-ranged and captive individuals that are available for the studies is limited. Preliminary results have been published in the final report of MINOS (National Park Regional Office for the Schleswig-Holstein Wadden Sea 2004). Initial results from field studies that exposed harbour porpoises to simulated operation noises of a 2-megawatt turbine indicate a clear avoidance of the source (Koschinski et al. 2003).

5 Harbour Seal

Harbour seals (*Phoca vitulina*) occur throughout the northern hemisphere. Critically important to the North Sea population are the pupping and nursery areas along the Wadden Sea coast. Harbour seals use the tidal flats for moulting and building up their lipid reserves during summer. They are optically well adapted both for above and underwater vision, and recent studies have shown that they also make intensive use of their whiskers as a tactile organ, which are susceptible to vibrations. Since they are also considered as having keen senses, they may be vulnerable to acoustic emissions, too.

Using telemetric methods, the MINOS research has found that feeding trips reach further than 50 km west of the haul-out sites in the Wadden Sea. Seals may spend several days feeding in the offshore waters, including anticipated wind farm areas (chapter 10). Thus, it cannot be ruled out that harbour seals may be displaced from their feeding grounds by the construction and operation of wind farms, possibly without adequate alternative locations. However, as discussed above, it has been suggested that the fishery ban in and near wind farms may lead to locally increased abundance and diversity of fishes and hence, wind farms may see increased activity of harbour seals. This effect could be enhanced by the enriched fauna on the hard substrates of piles amidst soft bottom areas which attract more fish. Given these possibilities, audiometric research is also conducted for seals, considering susceptibility to ramming noises and other vibrations. The telemetric results will indicate whether there are feeding grounds and/or associated haulout sites that coincide with planned wind farms. The MINOS follow-up has selected, in cooperation with Danish colleagues, additional appropriate haulout sites where seals can be equipped with the telemetry units, consisting of time-depth-temperature, posture and mouth gap recorders, compass, and satellite units (see chapter 10). This should also yield information on the seasonal as well as spatial variability.

In the wind farm site off Horns Rev, so far no indication of disturbance to seals has been found by the Danish scientists during construction and operation. However, the site has also never been known to be of particular significance for seals, used neither as a feeding ground nor for haulouts. On the other hand, there are findings from Swedish wind farms where disturbance by logistics was reflected in shortened haulout periods. Avoidance of an acoustic source emitting operational vibration was also observed for harbour seals (Koschinski et al. 2003).

6 Seabirds

The North and Baltic Seas provide habitat to numerous seabirds. The open sea is residence and feeding ground especially for divers and sea ducks, where about 25 species can be encountered, permanently or seasonally. Distribution patterns encountered in different years may vary considerably in time and space, thus requiring long time series. Hydrographic forces such as currents or eddies and fronts may affect the patterns significantly. Because of enriched nutrients and plankton, fronts may attract fish and piscivorous seabirds. Diving seabirds prefer the depth range between 10 and 20 metres, which is also the depth of interest to wind farmers because it balances proximity to the coast with a feasible work depth.

To assess the possibility of displacement of seabirds by wind farms, it is important to take into account not only the occurrence of seabirds but also the direct threat from the turbines and rotors (Dierschke et al. 2003). MINOS led the development of a technical index: the Wind Farm Sensitivity Index or WSI (Garthe and Hüppop 2004), the first of its kind. The strength of this index is the variety of factors considered; for example, manoeuvrability, flight altitude of birds, nocturnal flight activity, and vulnerability against ship and helicopter traffic. Two diver species, the common scoter, and four species of terns turned out to be most sensitive to wind farms. From this index, spatial distributions of significance have been derived, allowing for the identification of areas of potential conflicts, as well as of low conflict potential. In general, the potential for conflict decreases as one moves offshore. As with the previous species, any assessment needs to be made in light of other impacts. For instance, scoters are among the species most affected by oil pollution.

References

Bundesamt für Seeschifffahrt und Hydrographie (BSF - Federal Maritime and Hydrographic Agency) (2003) Standards for Environmental Impact Assessments of Offshore Wind Turbines in the Marine Environment. Hamburg and Rostock/Germany. Available at www.bsh.de

Dierschke V, Hüppop O, Garthe S (2003) Populationsbiologische Schwellen der Unzulässigkeit für Beeinträchtigungen der Meeresumwelt am Beispiel der in der deutschen Nord- und Ostsee vorkommenden Vogelarten. Seevögel Bd. 24 (3):61–72

Garthe S, Hüppop O (2004) Scaling possible adverse effects of marine wind farms on seabirds: developing and applying a vulnerability index. J Appl Ecol 41 (4):724–734

Kellermann A (1995) Das Trilaterale Monitoring- und Bewertungsprogramm. Ein Überwachungsprogramm für das Wattenmeer. Dt. Hydrogr. Zeitschr. Suppl. 5:165–169

Koschinski S, Culik B, Damsgaard Henriksen O, Tregenza N, Ellis G, Jansen C, Kathe G (2003) Behavioral reactions of free-ranging porpoises and seals to the noise of a simulated 2 MW windpower generator. Mar Ecol Progr Ser 265:263–273

Kutscher J, Stump N (2004) Environmental Research for Offshore Wind Energy Deployment in the German Exclusive Economic Zone of the North and Baltic Seas. European Seminar Offshore Wind Energy in Mediterranean and other European Seas. Available at http://www.owemes.it/interno_fr.htm

National Park Regional Office for the Schleswig-Holstein Wadden Sea (ed) (2004) Final Project Report MINOS "Marine warm-blooded animals in the North and Baltic Seas. Foundations for the Assessment of Offshore Windfarms" (Summary). Funding number at German Federal Environment Ministry: FKZ 0327520. Available at http://www.minos-info.org

Scheidat M, Siebert U (2003) Aktueller Wissensstand zur Bewertung von anthropogenen Einflüssen auf Schweinswale in der deutschen Nordsee. Seevögel Bd. 24 (3):50–60

Tougaard S (2003) Satellite tracking of harbour seals on Horns Rev: Use of the Horns Reef wind farm area and the North Sea. Report to Techwise A/S, March 2003. (available on request)

Part V: Public awareness and consultation

Chapter 15

Consultation and public involvement

Annika Wallaschek

German Federal Ministry for the Environment, Nature Conservation and Nuclear Safety, Bonn

Abstract

The amendment to the German Federal Nature Conservation Act (BNatSchG) in April 2002 established a national legal framework for the implementation of the European Birds Directive and the Habitats Directive beyond Territorial Waters into the German Exclusive Economic Zone (EEZ). Responsibility for the selection, designation and management of these marine NATURA 2000 areas lies with the Federal Government. In selecting the protected areas in accordance with Article 38(2) of the German Federal Nature Conservation Act, the Federal Environment Ministry consults relevant Federal ministries from the adjacent coastal Länder as well as the general public. The proposed areas were listed and mapped on the Internet under www.*HabitatMareNATURA2000*.de. The list was also published through newspaper advertisements and press releases in November 2003. These initiated the participation process of the public. Each citizen had the opportunity to submit a written comment or to contribute to the expert debate in public. For that purpose, three public hearing dates were held in the coastal Länder in December 2003. The federal ministries were involved in 2003 and once again involved from the beginning of 2004. The final proposals for NATURA 2000 sites (proposed Sites of Community Importance, pSCIs) were submitted to the European Commission in May 2004.

Consultation and public involvement

The German Federal Nature Conservation Act (*Bundesnaturschutzgesetz*, BNatSchG) prescribes the following steps in selecting proposals for NATURA 2000 sites in the EEZ:

- Participation of the Federal Ministries concerned;
- Arranging consultation with Federal States (Länder) which border on the EEZ (cf. Article 38 para. 2 fourth sentence BNatSchG); and
- Public involvement (cf. Article 38 para.2 third sentence BNatSchG).

For the initial coordination with the Federal ministries, the documents, i.e., maps with proposed NATURA 2000 sites and the accompanying standard data forms were forwarded to the relevant ministries in February 2003 with the option of submitting an opinion. The comments submitted by these ministries were replied to in writing by the Federal Ministry for the Environment, Nature Conservation and Nuclear Safety (Bundesministerium für Umwelt, Naturschutz und Reaktorsicherheit, BMU). An interministerial discussion also took place in May 2003.

At the beginning of June 2003, the above-mentioned documents were forwarded for consultation purposes to the Federal Coastal States (Länder) of Schleswig-Holstein, Lower Saxony, Mecklenburg-Western Pommerania, Hamburg and Bremen. These States also submitted comments, which were replied to by the BMU in writing, and discussions were held at the beginning of November 2003.

Public involvement began immediately after the conclusion of these Länder discussions. According to Article 38 paragraph 2 third sentence of BNatSchG, "Protected marine areas shall be selected with the involvement of the general public." No formalised procedure is prescribed in the German Federal Nature Conservation Act (unlike with the EIA or planning approval procedures); hence, there were no formal provisions (e.g., deadlines, procedural steps) that must be strictly adhered to. Nevertheless, the involvement of the public occurred on the basis of the usual formal procedural steps which included publication, disclosure of documents for public inspection, opportunity to submit comments, and hearings.

With regard to the publication requirement, on 12 November 2003 the BMU published a press release and placed advertisements in national, regional and local newspapers. Information was given about the project and the procedures, the proposed protected areas were listed, and the BfN (Federal Agency for Nature Conservation) homepage was referred to for further background information (see chapter 16). Furthermore, an opportunity to submit written comments was announced, as well as three dates for public hearings which were to be held in December 2003.

Direct contact was also made with groups representing public interests, conservation and user associations, and other potential stakeholders. There were in total 200 addressees. The complete nomination documents (maps with proposals for protected areas and standard data forms) with additional explanations were available for inspection for one month at the BMU in Berlin and at the BfN in Bonn.

The public hearings were held in Bremen on 1 December 2003, in Stralsund on 10 December 2003, and in Rendsburg on 11 December 2003. These hearings were attended by representatives of authorities at Federal, State, and municipal level, by conservation and user associations, by enterprises from the wind power, sand and gravel extraction, oil and gas exploration and production, and by private individuals. The total number of participants was 180. By the end of December 2003, 34 written comments had been submitted to the BMU, which first underwent a general evaluation, and then at a later date, were replied to individually.

While representatives of users were overwhelmingly critical or even dismissive of the proposed areas, nature conservation and environmental

Figure 1. Public hearing

associations generally welcomed the proposed areas, even though they called at the same time for further research and area extensions.

The evaluation by the BMU/BfN of the contributions to the hearings and of the written comments within the framework of public involvement was based on the selection criteria listed in the Habitats and Birds Directives. In line with the Directives, only arguments related to nature conservation were considered. In this respect, the evaluation did not give sufficient cause to make changes to the spatial extent of the NATURA 2000 sites proposed by the BMU/BfN in the EEZ. A small modification was made to the proposed site *Western Rønne Bank* (*Westliche Rönnebank*) on the basis of latest findings from exploratory dives by the BfN. Minor improvements were made to the standard data forms.

The ministries whose areas of competence were affected were once again involved from the beginning of 2004 with further opportunities to submit comments and input within the framework of interministerial discussions.

The final proposals for protected areas were submitted to the EU Commission in May 2004.

The latest maps with coordination data and further background information can be found at www.*HabitatMareNATURA2000*.de.

Chapter 16

Raising public awareness – *"HabitatMareNATURA2000.de"*

Katrin Wollny-Goerke

Hamburg

Abstract

Parallel to the selection and nomination process of the marine NATURA 2000 sites, a presentation concept for increasing public awareness was developed and implemented. This was the research project "developing and implementing a presentation concept to increase public awareness". Starting in late summer 2002, four different products for conveying information were designed:

- a booklet (initially in German, also in English at a later date)
- a video (German and English)
- a website (initially in German, also in English at a later date)
- an interactive CD-Rom.

These four products were developed almost at the same time, but they were published in two phases with slightly differing goals and target groups, depending on the status of the nomination procedure. Initially, the video, the website and the booklet were prepared. Their publication date was June 2003 at the joint ministerial meeting of the Helsinki and OSPAR Commissions with the Baltic and North Sea riparian states in Bremen. The interactive CD-Rom will be published in the second half of 2005.

All products confirm under the main title and name *HabitatMareNATURA2000*. The initial subheading, "Research for protection of the North and Baltic Seas", was updated in summer 2004 to "Research and protection for the North Sea and Baltic Sea", according to the nomination procedure of the protected areas.

Raising public awareness

Research projects over many years in the North and Baltic Seas delivered, at the end of the 1990s, clear indications of particularly ecologically valuable marine areas. However, it was only following the revision of the German Federal Nature Conservation Act (BNatSchG) in April 2002 that it became possible for Germany to also protect such areas in the Exclusive Economic Zone (EEZ) (i.e., seawards of the 12-mile zone, within the framework of the European protected sites network NATURA 2000). A detailed research programme, with around 20 projects, aimed to substantiate initial findings and to deliver specific data on the identification and delimitation of possible protected areas in open sea.

It quickly became apparent that the significance of special marine habitats and animal species offshore is mainly unknown to the general public. The areas concerned and the species relevant for the delineation of NATURA 2000 sites are mainly out of sight: they are found far off the coast, hidden below the sea's surface, and they are very rare or live reclusively. As a result, there is very little public awareness of the North and Baltic Sea ecosystems: neither of the species requiring special protection nor of the ecological effects of economic uses of the open sea. Even some representatives of user groups are not aware of the high ecological value of many marine areas. On this basis, the research project, "Developing and implementing a presentation concept to increase public awareness" started. Its goal was to provide extensive information to the public, to representatives of user groups, industry, authorities and science, and to give them fascinating insights into the marine world of the North and Baltic Seas. People should be convinced of the importance of protecting these unique marine animals and plants, and their habitats.

At the same time, the public was also informed of the in-depth research projects and their results on identifying and designating NATURA 2000 protected areas in the EEZ. The intention was to present the complex research to the lay public in an understandable and clear way.

Starting in late summer 2002, different products for conveying information were designed to be used in support of the NATURA 2000 nomination procedure:

- a booklet (initially in German, also in English at a later date)
- a video (German and English)
- a website (initially in German, also in English at a later date)
- an interactive CD-Rom.

At the beginning, it was important to find a name that connected all the products, allowed the identification of the issue and that could be easily remembered. The name should be used as an internet address, as a title for all products, and as a logo as well. It should be understandable to both German and English speakers. Our team decided on "*HabitatMareNATURA2000* – Research for Protection of the North and Baltic Seas".

The website can therefore be accessed at www.*HabitatMareNATURA2000*.de. The first part of the name serves as a heading and main title. The addition, "Research for Protection of the North and Baltic Seas" explains the main title while also supplementing it because it informs not only about the habitat but also about many research projects in the open sea concerning the NATURA 2000 sites. Furthermore, it was also important that this subtitle could be updated according to the nomination phase. This was the case in summer 2004 when the subtitle was changed to "Research and Protection for the North Sea and the Baltic Sea" following the *nomination* of the NATURA 2000 sites to the EU Commission.

HABITAT MARE NATURA 2000

Forschung und Schutz für Nord- und Ostsee

HABITAT MARE NATURA 2000

Research and Protection for the North Sea and the Baltic Sea

Figure 1. Name and Logo *HabitatMareNATURA2000*

The term "Habitat Mare" enables the German Federal Agency for Nature Conservation (BfN) to use it at a later date for other projects with a similar focus as the NATURA 2000 project, but then with another subtitle.

The development of the four products required comprehensive filming and investigation. Various and different shots of the research projects were taken by an experienced camera team during aerial counts, on several boat trips, and during subsequent laboratory work. The comprehensive collection of video materials from the BfN was used for the necessary underwater shots of the different habitat types and some animal and plant species. These materials were gathered as part of the research programme with tow cameras and hand-held underwater cameras. Film sequences of the relevant animal species were also compiled from archives or were re-shot.

The photo documentation for the projects was drawn in part from existing materials from research institutes and scientists or materials elaborated for this purpose. Some research works were also re-photographed. Most of the photos of animal species were compiled from various archives.

The participating marine research institutes provided fundamental support for both the photo and video documentation, and the presentation of the research findings.

The four products were developed almost at the same time, but they were published in two phases with slightly differing goals and target groups, depending on the status of the nomination procedure.

In the first phase, the video, the website, and the booklet were prepared. The date for publication was June 2003 at the joint ministerial meeting of the Helsinki and OSPAR Commissions (with the Baltic and North Sea riparian states) in Bremen.

The publication of the video at the ministerial meeting was very important because it made possible the effective presentation of Germany's position in the marine NATURA 2000 network at an international level.

The focus of the video was the illustration of the proposals for NATURA 2000 sites and their characteristics. The presentation of the research programme complemented this. The wealth of experience in the BfN, the renown of the involved German marine research institutes and their precise knowledge of the marine areas under discussion could be shown with the documentation of the detailed research projects. On this basis, the presentation of the proposed Marine Protected Areas (MPAs) outlined the species and habitat types relevant for the delineation of NATURA 2000 sites. The video of approximately 30 minutes showed many unique, rare and fascinating pictures. These made the viewers sensitive to the special importance of protecting these underwater habitats and threatened species.

Within an international context, Germany takes on with this project an important role in identifying and designating MPAs in Europe. Considering the focus on the North and Baltic Sea riparian states, this could even be considered a pioneering role at the time.

The booklet – under the *HabitatMareNATURA2000* Logo – entitled "Marine areas worth protecting: identification, research, conservation" provides brief information on and explains NATURA 2000 issues and the nomination procedure. The habitat types and animal species relevant for the delimitation of NATURA 2000 sites are listed and the detailed research programme is briefly referred to, as is the website. The booklet was primarily distributed as information for the public, for instance at the ministerial meeting in June 2003 and at the public hearings in December 2003. It was also sent to various establishments such as nature conservation associations, information centres, research institutes, user groups (e.g., wind park companies), authorities and offices in the field of nature conservation and industry. The booklet was sent as information aid, especially as a reference to the website. It can be ordered from the BfN.

The website was originally planned as a short version to be linked to the BfN homepage. However, during the development stage it became clear that this medium – in part due to its update options – should be given much greater significance than was originally planned. It was therefore developed as an independent website (www.*HabitatMareNATURA2000*.de) which is linked to the BMU and BfN homepages, but which can now be considered as the BfN's main information medium for marine NATURA 2000 issues. In order to indicate a clear relation to the BfN, there were some prerequisites, like a similar navigation structure to the BfN homepage, some equivalent navigation aids (e.g., the icons) and formatting of hyperlinks. The colours of the BfN logo should also determine the new website. Naturally, the BfN logo itself needed to be placed in the same position as in the BfN homepage. The website was also realised according to the special demands of the BITV (*Barrierefreie Informationstechnik – Verordnung*; Handicap Accessible Information Technology Regulation) which ensures its disability-friendly character.

The website is installed in German and in English.

The topics can be seen in figure 2. Starting with the comments on the legal situation of and the nomination procedure for Sites of Community Interest (SCIs) under the Habitats Directive and bird protection areas, both the research programme and the habitat types and animal species requiring protection are presented. The investigated areas that were later renamed as NATURA 2000 nominations were also dealt with under the heading *research programme*. A total of 10 maps of the North and Baltic Seas with the proposed and registered areas, and the respective species and habitat types relevant for delineation of NATURA 2000 sites can be found here. A very broad range of maps and research reports are available to users under *download*. There are links leading to participating marine

Figure 2. Website www.*HabitatMareNATURA2000*.de – Introduction page

research institutes, nature conservation authorities, and international bodies such as HELCOM and OSPAR. This range is also supplemented by information on dates and facts concerning the nomination procedure.

The main feature of the website is the visually striking presentation done with the aid of an exceptional number of photos, maps and even videos. Two short video clips present – each from the perspective of a research method – a nominated protected area from the North Sea and the Baltic Sea. These clips show the users rare species and underwater habitat types that laypersons – even those with an interest in science – would hardly see otherwise. The photos and video clips also create an emotional connection to this issue and awaken an interest in reading the detailed and rather dry technical text.

The website is regularly updated and supplemented in line with the status of the nomination procedure and latest scientific findings from the research projects. For example, since maps of the proposed and – later – of the nominated protected areas can be downloaded, licensing authorities for economic uses in the EEZ, nature conservation associations

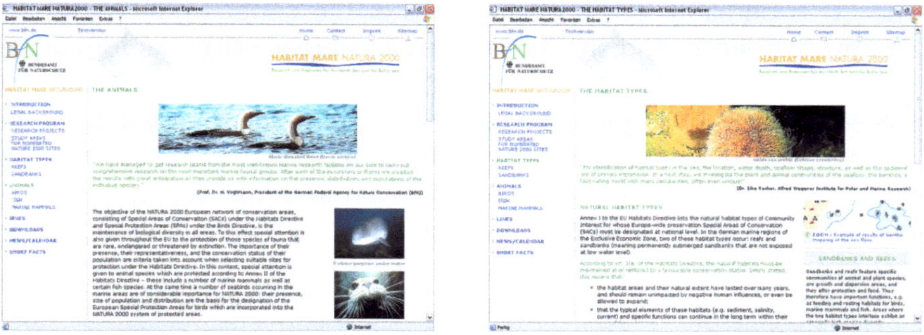

Figure 3. Website www.*HabitatMareNATURA2000*.de – Animals/Habitat types

and scientific institutes have the opportunity to refer to the latest maps whenever required. This also considerably eases the workload for the BfN as it can refer to the website when contacted with queries.

The fourth product is an interactive CD-Rom (published in 2006). This interactive CD-Rom aims to highlight the nominated protected areas and their special characteristics in particular in a simpler and more graphic way than the website. Many video clips introduce the user to the issue and illustrate the characteristics of the individual protected areas. Dozens of maps and figures, based on the actual data from the research projects, show the abundance and distribution of many threatened species of birds, marine mammals and fish, with the focus on the nominated protected areas. Many slide shows, several three- and two-dimensional animations with spoken texts and narrations were used to

- present the protected habitats as well as the animal species and their habitat requirements,
- show the special geological conditions of individual areas,
- provide an insight into the risks of exploitation and of economic uses of these areas of high ecological value, and
- give brief information about the legal status of the protected areas.

The presentations of the protected areas are placed in context with brief additional information on the North and Baltic Seas and on the areas of special ecological value that border the nominated protected areas. The CD-Rom also deals with the detailed research programme on marine NATURA 2000 issues, although at a less in-depth level than the website and the video.

The CD is primarily aimed at members of the general public with an interest in science, and should serve as a reference material for nature conservation associations, relevant authorities and enterprises. Although it is originally not designed as a computer application for use in a nature conservation information centre (since it is too detailed for this and contains too much text), it can be operated in nature conservation centres as long as a terminal is available and visitors can enjoy it without too much disturbance.

The booklet, video, and interactive CD-Rom are available from the BfN.

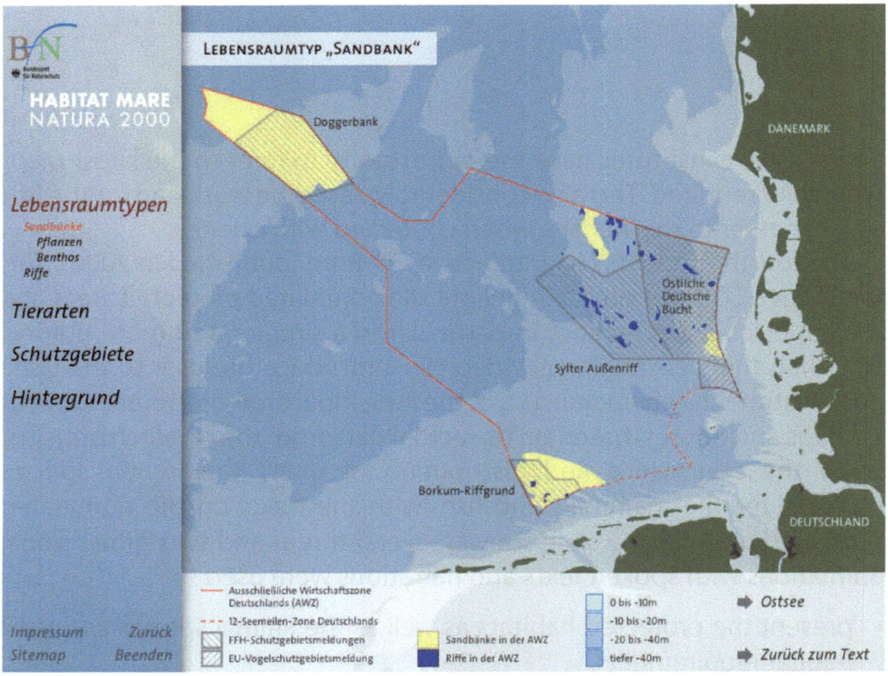

Figure 4. Interactive CD-Rom *HabitatMareNATURA2000* – Map of the threatened habitat types (sandbanks and reefs) in the EEZ of the North Sea

Acknowledgements

I would like to thank the following colleagues in my team for their engagement in this project: Achim Blöcker (ab medien); Helmke and Peter Kaufner (CINEDESIGN); Andreas Essenberger (essenberger neue medien); Kristin Popp (Popp Kommunikation); and Christian Terstegge. I also thank colleagues at the BfN and the scientists in the research projects for their support concerning actual data, photo and video material.

Annex

Addresses of Authors

Prof. Dr. Dieter Adelung
Leibniz – Institut für
Meereswissenschaften
Düsternbrooker Weg 20
24105 Kiel
Deutschland
Dadelung@ifm-geomar.de

Ina Backhausen
Bundesamt für Naturschutz
BfN Insel Vilm
18581 Putbus auf Rügen
Deutschland

Dr. Harald Benke
Deutsches Meeresmuseum
Katharinenberg 14/20
18439 Stralsund
Deutschland
harald.benke@meeresmuseum.de

Dieter Boedeker
Bundesamt für Naturschutz
BfN Insel Vilm
18581 Putbus auf Rügen
Deutschland
dieter.boedeker@bfn-vilm.de

Dr. Thomas Bosecke
Universität Rostock
Juristische Fakultät
Richard-Wagner-Strasse 31
18119 Rostock-Warnemünde
Deutschland
thomas.bosecke@uni-rostock.de

Prof. Dr. Detlef Czybulka
Universität Rostock
Juristische Fakultät
Richard-Wagner-Strasse 31
18119 Rostock-Warnemünde
Deutschland
detlef.czybulka@uni-rostock.de

Dr. Markus Diesing
Institut für Geowissenschaften
Christian-Albrechts-Universität zu Kiel
Olshausenstrasse 40-60
D-24098 Kiel
Deutschland
md@gpi.uni-kiel.de

Kai Eskildsen
Landesamt für den Nationalpark
Schleswig-Holsteinisches Wattenmeer
Schlossgarten 1
25832 Tönning
Deutschland
kai.eskildsen@nationalparkamt.de

Barbara Frank
Landesamt für den Nationalpark
Schleswig-Holsteinisches Wattenmeer
Schlossgarten 1
25832 Tönning
Deutschland
barbara.frank@nationalparkamt.de

Dr. Stefan Garthe
Forschungs- und Technologiezentrum
Westküste
Christian-Albrechts-Universität zu Kiel
Hafentörn 1
D-25761 Büsum
Deutschland
garthe@ftz-west.uni-kiel.de

Anita Gilles
Forschungs- und Technologiezentrum
Westküste
Christian-Albrechts-Universität zu Kiel
Hafentörn 1
D-25761 Büsum
Deutschland
gilles@ftz-west.uni-kiel.de

Dr. Fritz Gosselck
Institut für angewandte Ökologie GmbH
Alte Dorfstrasse 11
18184 Broderstorf
Deutschland
gosselck@ifaoe.de

Annette Groß
Bundesamt für Naturschutz
BfN Insel Vilm
18581 Putbus auf Rügen
Deutschland
annette.gross@bfn-vilm.de

Kathrin Heinicke
Bundesamt für Naturschutz
BfN Insel Vilm
18581 Putbus auf Rügen
Deutschland
kathrin.heinicke@bfn-vilm.de

Christopher Honnef
Deutsches Meeresmuseum
Katharinenberg 14/20
18439 Stralsund
Deutschland
christopher.honnef@meeresmuseum.de

Dr. Adolf Kellermann
International Council for the Exploration of the Sea (ICES)
H.C. Andersens Blvd 44–46
1553 Copenhagen
Denmark
adi@ices.dk

Andreas Klein
ARGUMENT GmbH
Stadtfeldkamp 30
24114 Kiel
Deutschland
Klein@Argument-Kiel.de

Dr. Jochen C. Krause
Bundesamt für Naturschutz
BfN Insel Vilm
18581 Putbus auf Rügen
Deutschland
jochen.krause@bfn-vilm.de

Nikolai Liebsch
Leibniz – Institut für Meereswissenschaften
Düsternbrooker Weg 20
24105 Kiel
Deutschland
nliebsch@ifm-geomar.de

Dr. Henning von Nordheim
Bundesamt für Naturschutz
BfN Insel Vilm
18581 Putbus auf Rügen
Deutschland
henning.von.nordheim@bfn-vilm.de

Dr. Eike Rachor
Alfred-Wegener-Institut für Polar- und Meeresforschung
27515 Bremerhaven
Deutschland
erachor@awi-bremerhaven.de

Dr. Meike Scheidat
Forschungs- und Technologiezentrum Westküste
Christian-Albrechts-Universität zu Kiel
Hafentörn 1
D-25761 Büsum
Deutschland
scheidat@ftz-west.uni-kiel.de

Dr. Klaus Schwarzer
Institut für Geowissenschaften
Christian-Albrechts-Universität zu Kiel
Olshausenstrasse 40–60
24098 Kiel
Deutschland
kls@gpi.uni-kiel.de

Dr. Ursula Siebert
Forschungs- und Technologiezentrum Westküste
Christian-Albrechts-Universität Kiel
Hafentörn 1
D-25761 Büsum
Deutschland
ursula.siebert@ftz-west.uni-kiel.de

Dr. Ralf Thiel
Deutsches Meeresmuseum
Katharinenberg 14/20
18439 Stralsund
Deutschland
ralf.thiel@meeresmuseum.de

Dr. Monica Verbeek
Aegis – marine environmental consultancy
Rua Adelino Mendes
Quinta do Choupal, Casa do Lago
2765-082 São Pedro do Estoril
Portugal
aegis@kqnet.pt

Ursula Verfuß
Deutsches Meeresmuseum
Katharinenberg 14/20
18439 Stralsund
Deutschland
ursula.verfuss@meeresmuseum.de

Annika Wallaschek
Bundesministerium für Umwelt,
Naturschutz und Reaktorsicherheit
12 06 29 Bonn
Deutschland
Annika.Wallaschek@bmu.bund.de

Prof. Dr. Rory Wilson
School of Biological Sciences
University of Wales
Swansea Singleton Park
Swansea SA2 899
Wales, UK
r.p.wilson@swansea.ac.uk

Katrin Wollny-Goerke
Kakenhaner Weg 170
22397 Hamburg
Deutschland
Dr.K.Goerke-K.Wollny-Goerke@t-online.de

Dr. Michael Zettler
Institut für Ostseeforschung Warnemünde
Seestrasse 15
18119 Rostock
Deutschland
michael.zettler@io-warnemuende.de